Introduction to Environmental Science

Lab Manual

Third Edition

Edited by Felicia P. Armstrong, PhD

Youngstown State University
Department of Physics, Astronomy, Geology
and Environmental Science

Many thanks to our contributors and reviewers:

Elizabeth Abraham
Isam Amin
Shadrack Ampomah
Breanna Beaver
Ray Beiersdorfer
Jennifer DeChellis-Houser
Jeffrey C. Dick

Taryn Hanna
Stefanie Hudzik
Alan Jacobs
Sarah E. Johnson
Leah Kaldy
Jonathan Kinney
Dan Kuzma

Mumtaz Laskar
Colleen E. McLean
Chantal Sakwe
Derek Scott
Kevin Summerville
Lawrenzo Yengwia

Cover image © Shutterstock.com

www.kendallhunt.com
Send all inquiries to:
4050 Westmark Drive
Dubuque, IA 52004-1840

Contents

Understanding the Scientific Method and Measuring

Objectives

1. Understand, order, and analyze the steps of the scientific method
2. Measure axes, mass, and volume using multiple instruments
3. Use significant figures, precision, and accuracy to describe measurements

Part 1A: The Scientific Method

Introduction

Science is a process, not a set of facts. Therefore, science cannot tell you what to think, but rather how to think, approach, and analyze problems. The process of scientific inquiry employs observation-based evidence (as opposed to opinions), gathered via experimental procedures that follow specific guidelines outlined by the **scientific method**. To study environmental science or any science the process of the scientific method must be understood and applied. The **scientific method** is the line of reasoning used by scientists to answer specific scientific questions. The process of employing the scientific method begins by making an **observation** and learning about that observed phenomena. Then a **hypothesis** or tentative explanation of the observation is developed. A hypothesis must be a testable statement. The validity of the prediction is tested by **experimentation** and further observations. The **results** from experimentation are interpreted and compared to previously known information then are used to support or reject the hypothesis. As such, science is not static, because new evidence can alter preexisting hypotheses. For this reason, it's important for scientists to **publish** their results in peer-reviewed journals, so other scientists and the public can read, interpret, and discuss the results. Hypotheses which have withstood rigorous testing can become theories and/or laws of science. A **scientific law** describes a process in the universe that has been tested by repeated experiments (the "how" something happens); it is usually represented by mathematical equations. A **scientific theory** has also been repeatedly tested and confirmed through experimentation; it explains the workings of a natural process (the "why" something happens). Both scientific theories and law must withstand rigorous testing by multiple researchers over a very long period of time. Theories integrate a combination of principles and hypothesis to explain and predict the workings of a natural process and are accepted by the scientific community.

Identifying the Steps of the Scientific Method

<u>Directions</u>: Organize the stack of cards into three color-coded groups, or scenarios. For each scenario, order the cards to represent the order of the scientific method (Observation, Hypothesis, Experimentation, Results, and Conclusion). Record the card numbers for each scenario into Table 1.1 below.

Table 1.1   Examples for Using the Scientific Method

	Scenario 1	Scenario 2	Scenario 3
Observations			
Hypothesis			
Experimentation			
Results			
Conclusion / Validation of Hypothesis			

Developing and Analyzing Hypotheses

<u>Directions</u>: Watch the video about beach pollution and record observations. Then, develop a hypothesis and consider how it would be tested.

VIDEO OBSERVATIONS (list four):

-
-
-
-

1. Write a hypothesis based on your observations. A hypothesis is a testable statement that is a proposed explanation of the observed event.

2. Describe how you would test your hypothesis.

3. What results would cause you to accept your hypothesis? Be specific.

4. What results would cause you to reject your hypothesis? Be specific.

5. Is a hypothesis ever proven to be *true*? Explain.

6. Explain the difference between a hypothesis and a theory.

7. Could your hypothesis ever become a theory? If so, explain how.

Part 1B: Measurements

Introduction

Precision and accuracy are important in any scientific experimentation as they beget reliable and repeatable results. Precision is a measure of repeatability while accuracy refers to how close the average value is to the true value. Although no measurement is exact, and precision and accuracy can vary depending on the measuring tool, rounding appropriately and using the correct amount of significant figures can be used to increase confidence in your findings.

Comparing Precision and Accuracy

<u>Directions</u>: Students from three labs (A, B, and C) measured the length of a wood block in centimeters. Measurements from the students are shown in Table 1.2. The lab instructor measured the wood block and found the actual length to be 3.51 cm. Based on this information, answer the questions below.

Table 1.2   Student Measurements

Repetition	Lab A (cm)	Lab B (cm)	Lab C (cm)
1	3.48	3.45	3.51
2	3.48	3.49	3.51
3	3.48	3.53	3.50
4	3.49	3.42	3.52

1. Which Lab had the most precise measurements? Circle: A / B / C

2. Which lab had the most accurate measurements? Circle: A / B / C

3. Which lab was the most imprecise and inaccurate? Circle: A   /   B   / C

Measuring Axes

<u>Directions</u>: Measure the axis (length, width, and height) and calculate the volumes of two objects in both English (inches) and Metric (centimeters) to three significant figures. Record your answers below.

Table 1.3   Object Dimensions in English and Metric units

		English Units (inches)	Metric Units (cm)
Object A: ________	Length		
	Width		
	Height		
	Volume (LxWxH)		
Object B: ________	Length		
	Width		
	Height		
	Volume (LxWxH)		

Determining Significant Figures

<u>Directions</u>: Record the correct number of significant figures next to each measurement.

3 = _______	314. = _______	3014 = _______
31 = _______	0.0314 = _______	301.3 = _______
314 = _______	0.314 = _______	301.40 = _______
3.14 = _______	3.140 = _______	0.30140 = _______
31.4 = _______	314.0 = _______	31400 = _______

Measuring Mass and Volume

<u>Directions</u>: With the balance, determine the mass of a writing utensil and a calculator in grams to four decimal places and record your answers in Table 1.4. To use the balance, turn it on, press the zero or tare button, place your pen/pencil inside, close the glass door, and record the mass. Determine the number of significant figures in each rounded number.

Table 1.4   Mass of Writing Utensil and Calculator

	Writing Utensil (g)	Calculator (g)
Original Measurement:		
Number of Significant Figures:		
Rounded to 3 Decimal Places:		
Number of Significant Figures:		
Rounded to 2 Decimal Places:		
Number of Significant Figures:		
Rounded to 1 Decimal Place:		
Number of Significant Figures:		

<u>Directions</u>: Place an empty beaker on the balance and record the mass. Using the markings on the side of the beaker, measure between 50 and 70 milliliters of water from the sink. Record the volume in column 1 of Table 1.5 below. Place the beaker on the balance and record the mass of water+beaker in Table 1.5, column 2. Subtract the empty beaker mass from the water+beaker mass and record the mass of water in Table 1.5, column 3. Finally, pour the water from the beaker into a graduated cylinder. Repeat these steps three times and total the volume and mass of water.

Weight of Beaker: _____________________ (g)

Table 1.5   Comparing Beakers and Graduated Cylinders

Repetition	COLUMN 1 **Measured Volume of Water in Beaker (ml)**	COLUMN 2 **Mass of Beaker + Water (g)**	COLUMN 3 **Mass of Water in Beaker (g)**
1			
2			
3			
Total			

What is the total volume of water in the graduated cylinder?: _________________________________

1. If 100 milliliters of water has a mass of 100 grams, what is the mass of 1 milliliter of water?

2. Based on the mass of water (Column 3), was the volume of water measured accurately in the beaker (Column 1)? Why or why not?

3. Based on the mass of water (Column 3), is the beaker or the graduated cylinder more accurate in measuring liquids? What evidence supports your decision?

Part B

Each group will be assigned an area to inventory the waste and recyclables and input the location on Table 2.2. From the assigned areas *hypothesize* on which area will have the <u>most recyclable</u> waste in the trash or lowest recycle rate (H1).

H1:

____

____

Justify your hypothesis H1 (why do you think this area will be like that):

____

____

    Hypothesize on which area will have the <u>highest amount of recyclables in the recycle bins</u> or the highest recycle rate.

H2:

____

____

Justify your hypothesis H2:

____

____

    NOTE: Do not remove the trash; just observe the type and amount of recyclables found in the assigned area. A minimum of **4** trash cans should be observed; 8 are preferred.

    Observation Area: _________________________________

    Number of trash cans observed: ________________________________

    Observation Notes:

____

____

Table 2.2   Inventory of **Trash Cans** from Various Areas around YSU Campus

Location	Group 1	Group 2	Group 3	Group 4	Group 5	Group 6
Aluminum Cans						
Plastic						
Cardboard						
Glass						
Mixed Paper						
How full?						

NOTE: Do not remove the recyclables; just observe the type and amount of each found in the assigned area. A minimum of **4** recycle bins should be observed; 10 are preferred.

Observation Area: _______________________

Number of recycle bins observed: _______________________

Observation Notes:

____

____

Table 2.3   Inventory of **Recycle Bins** from Various Areas around YSU Campus

	Group 1	Group 2	Group 3	Group 4	Group 5	Group 6
Location						
Aluminum Cans						
Plastic						
Cardboard						
Glass						
Mixed Paper						
How full?						

Answer and discuss the following:

1. Compare the different areas on campus with respect to the amount and type of recyclables being thrown away (Tables 2.2 and 2.3). What general factors do you think contribute to items being recycled or thrown in the trash (name at least 3)?

2. Did you accept or reject hypothesis H1? Why?

3. Which area examined in class had the <u>most</u> recyclables in the trash while at the same time had the <u>least</u> in the recycle bins (area where recycle bins were not being used)? What reasons do you propose contributed to recyclables being thrown away and not recycled <u>in this area</u>?

4. Did you accept or reject your hypothesis H2?

5. Which area examined in class had the <u>least</u> recyclables in the trash and at the same time had the <u>most</u> in the recycle bins? What reasons do you propose contributed to higher recycling rates <u>in this area</u>?

6. What other items, <u>not</u> examined in this lab, can be recycled (name at least 2)?

7. What items can you purchase that are made from recycled products (name at least 3)?

8. After reviewing this information, does your opinion of recycling change, improve or alter? Explain.

9. How would you persuade someone to start recycling?

Minerals and Their Uses

Objectives

1. Describe a mineral's properties and characteristics through various tests and observations.
2. Identify products made from minerals.

Materials Needed

Mineral samples
Penny
Porcelain scratch plate
Steel nail
Steel file
Glass plate
Magnet
Dilute hydrochloric acid
Product samples

Introduction

Minerals are defined as naturally occurring, inorganic, solids with a definite chemical composition and a regular, internal crystalline structure. The composition of a mineral can be defined by its chemical formula. A chemical formula identifies each element contained in the mineral using the element's chemical symbol. Compositions vary greatly: for example, halite has the chemical composition of $NaCl$, whereas topaz has a chemical composition of $Al_2SiO_4(OH,F)_2$. These are examples of chemical compounds which are defined as two or more different chemical elements in a fixed ratio. Most minerals are chemical compounds, whereas a small number (e.g., sulfur, copper, gold) are elements.

Approximately 3,000 minerals exist in nature. To identify them we use a series of test to determine the minerals physical properties. The color, streak, luster, hardness, reaction to acid, magnetism, and transparency are some physical attributes that can be used to identify minerals.

1. **Color** – Color is one of the most obvious descriptors of a mineral. There are many minerals that have distinctive color; but there are also many that come in a variety of colors. A mineral's color is a result of the mineral's chemical composition, impurities that may be present, or flaws or damage in the internal structure. Even though color is the easiest physical property to determine, it should be used with other diagnostic tools for mineral identification.

2. **Streak** – Streak is the color of a mineral when it is powdered. Streak can be determined for any mineral by dragging an edge of the mineral across a porcelain plate or streak plate. Some minerals have streaks very different from their color. For example, the grey mineral

hematite leaves a red-brown streak. To make a streak, set the porcelain plate on a flat surface, such as a table top. Never streak a mineral while holding the plate in your hand. Hold the plate on the table with one hand and carefully drag the mineral across the plate while applying a little bit of downward pressure. The mark left by the mineral is the streak color. If the streak is white, you may not see it against the porcelain plate; you will need to look closely.

3. **Luster** – Luster of a mineral is the way a mineral reflects light or its sheen. The different types of luster are **metallic**—having the look of polished metal; **submetallic**—having the look of a metal that is dulled by weathering or corrosion; **nonmetallic**—not looking like metal at all. Nonmetallic luster can further be described as a subtype: **glassy**—having the look of glass; **pearly**—having the iridescent look of mother of pearl (though usually just barely); **waxy**—greasy/oily, silky, resinous; **dull**—earthy or nonreflective surface.

4. **Hardness** – Hardness refers to how resistant a mineral is to being scratched or its resistance to abrasion. Some minerals can be scratched easily with a fingernail, others cannot. Diamond is the hardest mineral because it can scratch all others. Talc is one of the softest minerals; nearly every other mineral can scratch it surface. Hardness is measured by comparing it to minerals of known hardness. **Mohs Hardness Scale** is used to compare the hardness of the unknown mineral to known minerals (Table 4.1). The scale runs from soft or 1 such as the mineral talc to diamond at 10, the hardest mineral. Scratch the surface of the unknown mineral with various known hardness items (e.g., penny, steel nail, etc.) to determine the relative hardness.

Table 4.1   The Mohs Hardness Scale

Hardness	Index Mineral	Reference Test
1	Talc	
2	Gypsum	Fingernail will scratch the mineral (hardness <2.5)
3	Calcite	Penny will scratch the mineral (hardness 3-4)
4	Fluorite	
5	Apatite	Steel nail will scratch the mineral (hardness 5)
6	Orthoclase	Mineral scratches glass (hardness >5.5)
7	Quartz	Steel file will scratch the mineral (hardness >7)
8	Topaz	
9	Corundum	
10	Diamond	

5. **Reaction to acid** – Although this is actually a chemical property, it is very useful because only a few minerals react to dilute acid. Minerals containing the carbonate ion (CO_3^{2-}) effervesce or fizz when a drop of dilute hydrochloric acid (HCl) is added. In the reaction, carbon dioxide (CO_2) is released from the mineral and bubbles through the acid creating the fizz. To test the unknown mineral, place the mineral on a paper towel and place 1-2 drops of acid on the mineral. Wipe the acid from the mineral after recording the observations.

   Caution—Be careful not to get HCl on your skin or in your eyes because it can cause irritation. If you get HCl on your skin or in your eyes, immediately flush with water and then notify your instructor that you have got HCl in your eyes. Be careful not to get HCl on your clothes because it can make holes in them.

6. **Magnetism** – A few minerals are attracted to a magnet or are themselves capable of acting as magnets. Although not too prevalent in minerals, it is a useful property in making identification. To test the mineral place a magnet on the mineral; if the magnet sticks it is magnetic, if not then it's not magnetic.

7. **Transparency** – The way light passes through a mineral determines the transparency. More light passes through **transparent** minerals, **translucent** minerals let partial light pass through, and **opaque** minerals do not let any light through. If an object can be seen with a clear outline through the mineral then that mineral is transparent. If an object cannot be seen through, but light is transmitting light then that mineral is translucent. An opaque mineral is one that does not transmit light.

Mineral Extraction and the Environment

According to the Mineral Information Institute, 37,687 pounds of new mineral are needed for every person in the United State to make things we use every day (Table 4.2). Minerals have been extracted from the earth for hundreds of years using a variety of techniques. The two basic types of extractions are **surface** (open pit, strip) and **subsurface** (deep or underground) methods. Surface mining is the most common form of mining in the United States with about 85% of minerals extracted using this method. In the process of mineral extraction, the surface vegetation, dirt and, if necessary, bedrock are removed to expose the buried minerals. Therefore the earth's surface remains exposed during the process of surface mining, where drilling and blasting are used to break up and remove the product. Minerals that are surface mined include halite, gypsum, and sand.

Subsurface or underground mining consists of digging tunnels or shafts into the earth to reach the deposits of minerals or ores. The mineral or ore and waste rock are brought to the surface through the tunnels. Subsurface mining is classified by the type of tunnels or shafts that are used. Drift mining uses horizontal access shafts, slope mining uses diagonally sloping access shafts, and shaft mining uses vertical access shafts. Minerals that are subsurface mined include graphite, halite, diamond, and pyrite.

There are many **environmental effects** as a result of mining activities. Erosion, sinkholes, loss of biodiversity, contamination of soil, groundwater, and surface water by chemicals used in the mining process including acid mine drainage are some of these effects. As a result, mining companies are required to follow stringent environmental and reclamation codes to reduce the environmental impact and avoid impacts on human health. The Surface Mining Control and Reclamation Act (1977), Resource Conservation and Recovery Act (1976), and the Comprehensive Environmental Response, Compensation, and Liability Act, or Superfund Act (1986) try to address some of the environmental impacts due to mining activities. In addition, Clean Water Act (1972) and the Clean Air Act (1970) helped to manage some of the air and water pollutants. Another act that affects mining in the United States is the General 1872 Mining Act. This now controversial act, which was originally designed to encourage settlement of the West, allows mining companies to purchase land for $2.50 or $5.00 per acre and has no statements of environmental reclamation. Provisions of the 1872 Mining Law were changed with the implementation of the 1976 Federal Land Policy Management Act (FLPMA), which revised some of the surface uses allowed on mining claims on public lands, and required more financial responsibility, more reclamation, and detailed mining plan of operations prior to disturbing the mining site. In 2009, the Hardrock Mining Reform Act was introduced to increase the fees and royalties associated with mining and install more rigorous environmental controls and reclamation actions.

Table 4.2   Amount of New Minerals Needed Per Person in the United States to Provide for the Things We Use Every Day (www.mii.org)

Mineral Needed	Lbs./year /person	Lbs./ lifetime	Uses
Stone	8,343	1.11 million	Used to make roads, buildings, bridges, landscaping and for numerous chemical and construction uses.
Sand, gravel, and cement	5,937		Used to make concrete, asphalt, roads, blocks, and bricks.
Cement only	530	41,181	Used to make roads, sidewalks, bridges, buildings, schools, and houses.
Iron ore	187	14,530	Used to make steel (buildings, cars, trucks, planes, and trains), other construction, containers.
Salt	409	31,779	Used in various chemicals, highway deicing, food, and agriculture.
Phosphate rock	195	15,152	Used to make fertilizers to grow food, bricks, cement, and paper.
Aluminum (bauxite)	52	4,040	Used to make buildings, beverage containers, autos, and appliances.
Copper	12	932	Used in buildings, electrical and electronic parts, plumbing, and transportation.
Lead	10	777	75% used in transportations batteries, electrical, communications and TV screens.
Zinc	7	544	Used to make metals rust resistant, various metals and alloys, paint, rubber, skin creams, and health and nutrition.

Clarence R. Smith Mineral Museum

Objectives

1. Explore the Mineral Museum.
2. Learn where some different minerals have been found.

Introduction

Clarence R. Smith was born and raised in Youngstown, Ohio. He started his mineral collection in 1959 when he spent a winter in Tucson, Arizona. In 1962, Mr. Smith opened a mineral shop and in 1970 after Mr. Smith passed away the minerals were moved to Adamas Jewelry and Gifts in Boardman, Ohio. The collection was a gift to Youngstown State University, Department of Geological & Environmental Sciences and in 2001 the Mineral Museum was opened on the ground floor of Moser Hall. Besides containing many world-class mineral specimens the collection also contains many unique specimens that are only available in a few other museums. Currently the museum serves as an educational resource for YSU faculty and students, local teachers, and the community.

Answer the following questions while touring the museum:

1. A variety of this mineral is named after a popular summertime fruit because it's pink on the inside with a green rim. _______________________________

2. Black quartz is called _______________ quartz.

3. Blue smithsonite contains the metal _______________

4. Cut and polished pieces of rock are called _______________________

5. Kidney Ore is an ore of _______________

6. Minerals fluoresce when they are exposed to _______________________ light.

7. Minerals will fluoresce only if they contain specific atoms known as _______________ s.

8. Ohio is famous for all of the following minerals except (circle the correct answer)

    amethyst;   calcite;   celestite;   fluorite;   halite;   pyrite.

9. One of the chemical elements in this purple mineral is an active ingredient in toothpaste

    ____

10. What two elements make up the composition of meteorites on display in the museum? _________________ and _______________.

11. Purple quartz is called _____________________________

12. The "Pine Cone Calcite" specimens in the Clarence R. Smith Mineral Museum come from the country of _________________

13. The formula for realgar and orpiment is _________________

14. The fossil that looks like a sea lily is called a _____________________

15. The fossilized dinosaur feces is called a _______________________

16. The large concretions in the museum are made out of _______________

17. The light pink gemstone in the gemstone cabinet is called _____________________

18. The log cabin is made from fossilized _________________ stems.

19. The yellow mineral in the color cabinet is _____________________

20. These three metals form minerals as pure (native) elements C _____________; G _____________; S _________________.

21. This country is famous for large specimens of amethyst. _____________________

22. This green mineral has the formula $Ca_2 Al_2 Si_3 O_{10} (OH)_2.$ _____________________

23. This mineral comes in green, pink, white, and yellow and crystallizes like a brush, a cathedral, a pine cone, or poker chips. _________________

24. This white mineral has the formula $Na Ca_2 Si_3 O_8 (OH)$ _____________________

25. I was born in the month of _____________________ and my birthstone is _____________________

BONUS: Which mineral in the museum do you think has the most value?

Topographic Maps and Global Positioning System

Objectives

1. Understand how to read and use a topographic map and contour lines.
2. Learn how to use a GPS receiver.
3. Plan a course of action based on topographic maps.

Materials Needed:

Topographic maps
Rulers and string
GPS receivers
YSU Campus Maps

Part A: Topographic Maps

Introduction

Topographic maps are maps that portray the Earth's three-dimensional surface on a two-dimensional map. They contain contour lines to represent the shape and elevation of the land. **Contour lines** are lines drawn on a map connecting points at equal elevation above sea level. The numbers written on contour lines or **index contours**, indicate the elevation above sea level. Every fifth contour line is labeled with elevation and shown by a bolder line on the map. The spacing of the contour lines varies with the change in elevation. If contour lines close together it indicates a more rapid change in elevation; widely spaced contour lines indicate a flat terrain or one with little change in elevation. The **contour interval** is the difference in elevation between two successive contour lines. The contour interval is determined by the **relief** of the mapped area. Areas that have large change in elevation or high relief will have larger contour intervals to prevent the map from having too many contour lines, which could make the map difficult to read. The elevation of unlabeled lines can be determined by the contour interval noted on the legend at the bottom of the map. To determine the elevation of an unlabeled line, count the number of contour lines from a labeled line and multiply by the contour interval. The **slope or gradient** describes the steepness of a change in elevation; the greater the slope the greater the steepness. To determine the slope, determine the change in elevation and divide it by the distance in feet of the area in question.

To identify what part of the Earth a map represents you need to identify the latitude and longitude (Figure 6.1). **Latitude** line runs horizontally spaced about 69 miles apart. Degrees latitude are numbered from 0° (equator) to 90° north and south. **Longitude** lines are also known as meridians. They stretch from the north to the south pole and are 69 miles apart at the equator. Zero degrees longitude is located at Greenwich, England (**prime meridian**) and lines continue 180° east and 180° west. To locate points on the Earth's surface, degree latitude and longitude have been divided into minutes (') and seconds ("). There are 60 minutes in each degree and each minute has 60 seconds.

The best known United States Geological Survey (USGS) maps are the 7.5-minute quadrangles. The 7.5-minute quadrangle map covers a four-sided area of 7.5 minutes of latitude and 7.5 minutes of longitude. The United States has been systematically divided into precisely measured quadrangles, and adjacent maps can be combined to form a single large map. The **fractional scale** of a topographic map can be found in 1:24,000, 1:100,000 and 1:250,000 as well as various other scales. A fractional scale of 1:100,000 means that 1 inch on the map represents 100,000 inches (1.6 miles) on Earth's surface.

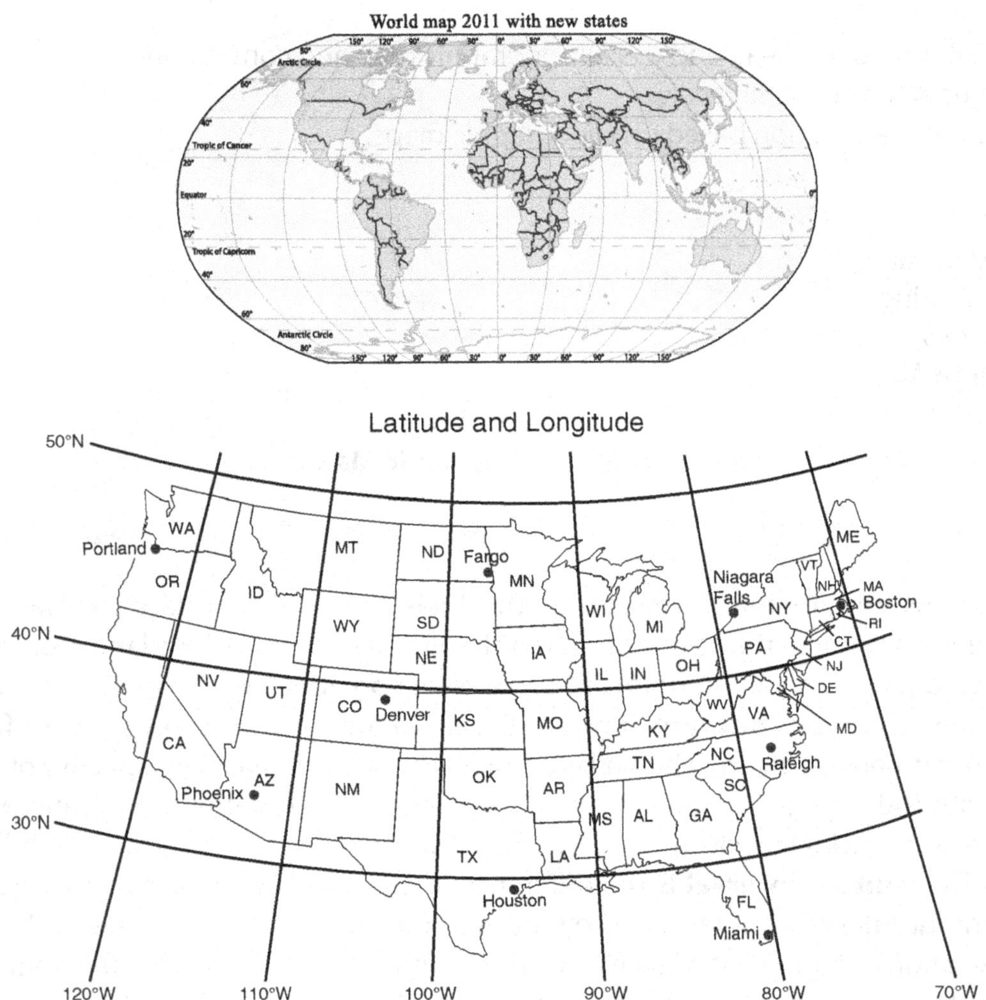

Figure 6.1   World map showing latitude (horizontal lines) and longitude (vertical lines). The equator is 0° latitude while the prime meridian is 0° longitude (top). Map of the United States showing latitude and longitude lines. The longitude lines are west of the prime meridian (0 degrees) therefore, numbers (degrees) get larger as your go west. Latitude lines are north of the equator (0 degrees) therefore, the numbers (degrees) increase as you go north (bottom).

Part B: Global Positioning System

Global Positioning System (GPS) is a U.S. based satellite radio navigation system that provides reliable positioning for anyone with a GPS receiver. The GPS is made up of three parts: 24 satellites orbiting the Earth, control monitoring stations on the Earth, and the GPS receivers. GPS satellites send out signals from space that is picked up by the GPS receivers. Each receiver then provides location (latitude and longitude) and elevation. There are a variety of applications for GPS including aviation, road and highway planning, precision agriculture, geology, and many environmental applications from location of contaminated sites or spills to location of breeding sites.

Part A Exercise:

Carefully study the Youngstown, Ohio topographic quadrangle map and the publication on USGS Topographic Map Symbols that have been provided. Then answer the following questions:

1. What is the scale of the map? ____

2. What is the contour interval of the map? _______________________________________

3. How many inches on the Earth does one inch on the map represent? _______________

4. Using the locations given below, identify 3 human made and 3 natural features.
   A.   Human Made or Altered Features (e.g., highways, public buildings, cultivated land)

   Location   Feature

   N 41° 00′ 40″ W 80° 41′ 40″   ____

   N 41° 03′ 15″ W 80° 39′ 20″   ____

   N 41° 06′ 30″ W 80° 37′ 40″   ____

   B.   Natural (low impacted) Features (e.g., wetland areas, forest)

   Location   Feature

   N 41° 05′ 45″ W 80° 40′ 30″   ____

   N 41° 00′ 55″ W 80° 43′ 00″   ____

   N 41° 01′ 45″ W 80° 39′10″   ____

5. Identify following features:

   a) The lowest elevation of the map ____

   b) The highest elevation of the map ____

   c) What is the elevation of Moser Hall (approximate)? ____________________________

6. What are the names of the adjoining topographic quadrangles?

North ________________________________   Northeast ________________________________

East ________________________________   Southeast ________________________________

South _______________________________   Southwest ______________________________

West ________________________________   Northwest ______________________________

7. Why is the map referred to as a 7.5 minute series topographic map?

8. Using geographic landmarks, Identify where the steepest slopes and the flattest terrain can be found.

9. Background Information
    Use a string and follow the rivers pathway from where it starts on the map to where it leaves the map. Take that length of string, and measure it using the correct scale found in the legend at the bottom of the map.
    Which direction is the Mahoning River flowing? ______________________________________
    How can you tell? ____

10. What is the length of the Mahoning River? Feet: ______________ miles: _______________

11. What is the elevation change or relief of the Mahoning River? ____________________ feet.

12. What is the slope or gradient of the Mahoning River? ________________________ feet/mile
    (Hint: Divide answer 11 by 10)

13. The length of Mill Creek?   feet:_______________________miles:_____________________

14. The change in elevation of Mill Creek? _____________________________ feet.

15. The gradient of Mill Creek in feet per mile? ________________________________ feet/mile.

16. How does Mill Creek *gradients* compare to the Mahoning River?

____

____

____

____

____

____

Observations/Problem:

There was a chemical spill at 8:22 am at Furnace Road and the north east side of Bears Den Road (approx. N41°4'55", W80°41'55"). This chemical has been known to cause harm to aquatic organisms. Officials and the remediation team need to know which surface water, Ax Factory Run or Bears Den Run, would be the first to come into contact with the chemical contaminant. If the team sets up barriers in the wrong area, serious damage will result to the aquatic organisms in the river.

Write a testable hypothesis to describe the direction of flow of the chemical contaminant based on the topographical map information. **Support or explain** your reasoning for your hypothesis.

H:

Support:

How could you test your hypothesis?

What **other** factors would you have to consider when working on this chemical spill?

Part B: GPS

As a group you will explore the Youngstown State University campus using a GPS receiver. Each group will sign out a GPS receiver and travel to the sites listed below. At each site you will either record the latitude, longitude, and elevation, or location description of the coordinates. Be sure to write clearly and identify whether coordinates are N, S, E, or W.

Review with your instructor how to use the GPS units before leaving. When finding coordinates you need to be at the 15 ft range (or as close as possible to the 15 ft range) to find the location.

Using the YSU map, start from the plaza doors of Moser Hall; head southeast to *reach for the stars* to the coordinates: N 41° 6′ 16.4″ W 80° 38′ 50.0″ elevation approximately 915 feet.

Site 1 Description: ____

We need to take in some *culture* so from Site 1 head northeast toward Wick Avenue for **Site 2** at N 41° 6′ 19.4″ W 80° 38′ 44.5″ elevation approximately 950 feet.

Site 2 Description: ____

This circle drive is considered the front entrance to the YSU campus. On this street is the Coffelt Hall where you will find the graduate school. Give the elevation of the YSU sign.

Site 3 Elevation: __________________ (don't forget units)

Bump, set, spike!. From Site 3 head northwest to N 41° 6′ 25.5″ W 80° 38′ 51.2″ elevation approximately 954 feet. Don't go too far, **Site 4** is before the basketball courts but after Sweeney Hall (Note: In winter the area could be covered in snow).

Site 4 Description: ____

Flying in the wind you continue on northwest to find **Site 5** located at N 41° 6′ 27.4″ W 80° 38′ 58.0″ elevation approximately 970 feet.

Site 5 Description: ____

From Site 5 head south to **Site 6** and give the coordinates to the "Rock".
Accuracy __________________ ft

Site 6 Latitude N__________________ Longitude W__________________
Elevation __________________

Life is a *work of art* and it is time to head back to our starting location. **Site 7** is southwest of the rock at N 41° 6′ 20.9″ W 80° 38′ 54.9″, elevation approximately 930 feet.

Site 7 Description: ____

Photosynthesis and Plant Cells

Objectives

1. Perform an experiment to describe the effect of light quality and quantity on the rate of photosynthesis.
2. Use paper chromatography to separate the photosynthetic pigments in a Spinach leaf extract.

Materials:

Anacharis plant	Light source
Microscope	Aluminum foil
Microscope slide and cover slip	Straw
6 sample vials and caps	2 Strips of chromatography paper
Large test tube rack	Glass rod
Bromothymol blue	Spinach leaves
MicroLab	Chromatography solvent
6 spectrometry vials and caps	

Introduction

All plants (both terrestrial plants and aquatic including phytoplankton) are the most important carbon sinks, taking up vast quantities of CO_2 through the process of ***photosynthesis***. To a lesser extent, atmospheric CO_2 can also be dissolved directly into ocean waters and thereby be removed from the atmosphere. While plants as well as animals release CO_2 through the process of respiration, the amount of CO_2 taken up by plants through photosynthesis and released through respiration approximately balances out annually on a global basis. Thus, the CO_2 released from human activities is truly the "extra" CO_2.

Photosynthesis is the process of converting light energy to chemical energy (glucose) and in the process oxygen (O_2) is released and CO_2 is locked up in glucose and plant tissue. The second part of the lab you will examine the aquatic plant Anacharis. Like any complex cell, it has the basic organelles (mitochondria, chloroplast, etc.). The nucleus with DNA will not be visible, although the cell wall and chloroplasts will be very obvious. The **chloroplasts** are the site for photosynthesis in the cell; they contain **chlorophyll**, which traps sunlight and thus enables the cell to combine CO_2 and water to make glucose ($C_6H_{12}O_6$).

Anacharis, also known as the Brazilian Waterweed is a common "exotic" aquarium plant. Anacharis is an excellent plant for our studies of photosynthesis and cells because it is easy to grow, and readily available. The leaves are only a few cells thick, so they will be easy for us to observe under the microscope to look at cells and cell parts.

Part A: Photosynthesis

Photosynthesis: $H_2O + CO_2 + \text{Light energy} \rightarrow C_6H_{12}O_6 + O_2$

Water + Carbon Dioxide + Light Energy $\rightarrow$ Glucose + Molecular Oxygen

Examining the equation above, we see as photosynthesis progresses, carbon dioxide is consumed and oxygen is liberated.

Respiration: $C_6H_{12}O_6 + O_2 \rightarrow CO_2 + H_2O + \text{energy}$

Glucose + Molecular Oxygen $\rightarrow$ Carbon dioxide + Water + Energy

In plants when there is no light to provide energy, *respiration* occurs. During respiration carbon dioxide (CO_2) is liberated as glucose is broken down and energy is released.

Plants produce pigments, such as chlorophyll, that enable plants to utilize sunlight for energy. There are two different types of chlorophyll, chlorophyll a and chlorophyll b, both of which can be found inside plant cells. These are produced in the leaves in large quantities during the spring and summer giving leaves their green color. Pigments other than chlorophyll can be found in plants and are quite useful for collecting the energy in the various wavelengths of light that chlorophyll is incapable of accessing. These pigments (e.g. yellow carotenes and yellow-orange xanthophyll) can be teased apart and seen separately from one another in a process called **paper chromatography**.

Chlorophyll is important to plants to make food using sunlight. During the summer and spring there is lots of light (energy) and the plant makes lots of chlorophyll. In the **fall** when the length of daylight changes and temperatures cool, leaves stop producing chlorophyll. It take energy for the plant to make the chlorophyll, so the plants break down the chlorophyll and move it out of their leaves before they fall. The green color dissappears so that the other pigments, yellow to red, become visible.

There are three sections to this lab:

Part A: Photosynthesis
Part B: Examination of Plant Cells
Part C: Separating Photosynthetic Pigments

After each section you will record your results and answer questions that correspond to that section of the lab.

Formulate a hypothesis for which water solution (1/8 tsp, 1/4 tsp, or controls) will photosynthesize most rapidly and why.

H:

Methods

1. Use the hole punch to create leaf discs. Be sure to avoid large leaf veins and any areas of the leaf that are damaged. You should have 40 leaf discs that are green and healthy looking.
2. Add 1/8 teaspoon of bicarbonate to 300 mL of water. Mix well. Add 1 drop of liquid dishwashing soap to the mixture. There should be little to no suds. To another 300 mL of water add one drop of liquid dishwashing soap.

3. Place 10 discs in the plastic syringe. Replace the plunger and depress, but do not crush the leaf discs.
4. Insert the tip of the plastic syringe into the bicarbonate solution and draw up about 20-30 mL. Be sure all leaf discs are floating and as much as possible, not sticking together.
5. When all discs are floating, cover the end of the plastic syringe with your finger and pull the plunger to for 15-20 seconds to create a vacuum. This step will need to be repeated until all leaf discs have sunk to the lower end of the syringe.
6. Pour 50 mL of the bicarbonate solution into a beaker. Add the contents of the syringe to the beaker. Be sure no discs are sticking together or to the inside of the beaker. They should all sink to the bottom of the beaker.
7. Quickly measure 30 cm from the beaker and place the light in front of the beaker. Begin recording the time. Record the time in minutes and record when discs begin to float on Table 7.1.
8. While waiting for discs to float, add 10 discs to the plain water and liquid dishwashing soap solution. This will act as the control.
9. Repeat 1-7 with 1/4 teaspoon of bicarbonate and 300 mL of water and a second control.
10. Record results in Table 7.1.

Table 7.1   Number of leaf disc floating with time in different solutions

$\frac{1}{8}$ tsp Bicarbonate Solution		Control 1		¼ tsp Bicarbonate Solution		Control 2	
Minutes	**No. of Leaf disc floating**	**Minutes**	**No. of Leaf disc floating**	**Minutes**	**No. of Leaf disc floating**	**Minutes**	**No. of Leaf disc floating**
1		1		1		1	
2		2		2		2	
3		3		3		3	
4		4		4		4	
5		5		5		5	
6		6		6		6	
7		7		7		7	
8		8		8		8	
9		9		9		9	
10		10		10		10	
11		11		11		11	
12		12		12		12	
13		13		13		13	
14		14		14		14	

15		15		15		15	
16		16		16		16	
17		17		17		17	
18		18		18		18	
19		19		19		19	
20		20		20		20	
21		21		21		21	
22		22		22		22	
23		23		23		23	
24		24		24		24	
25		25		25		25	
26		26		26		26	
27		27		27		27	
28		28		28		28	
29		29		29		29	
30		30		30		30	

1. Graph your results in Figure 7.1 for the 1/8 tsp Bicarbonate Solution with time and Control 1 and Figure 7.2 with the 1/4 tsp sodium bicarbonate and control 2. Be sure to label your x and y axes appropriately and add a legend to each graph to identify which results are the control and sodium bicarbonate solution.

Figure 7.1   Result of leaf photosynthesis measured through from the number of leaf disc floating in a solution containing 1/8 tsp sodium bicarbonate and control 1.

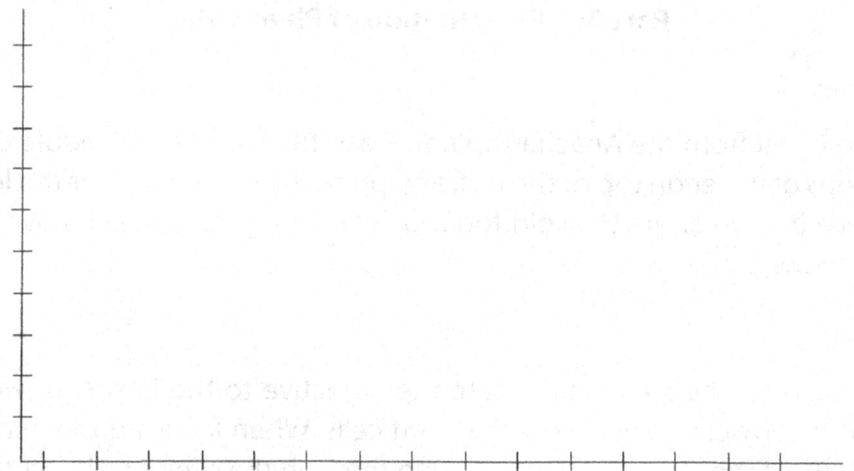

Figure 7.2   Result of leaf photosynthesis measured through from the number of leaf disc floating over time for a solution containing 1/4 tsp sodium bicarbonate and control 2.

2. What does the bicarbonate provide the leaf disc so it can perform photosynthesis?

3. What was the rate of photosynthesis (how many leaves floated per minute) for each sodium bicarbonate solution?

4. What purpose does the control serve?

5. Was your hypothesis accepted or rejected? Why or why not?

Part B: Examination of Plant Cells

Preparing the Slide:

Gently take a single leaf from the Anacharis plant. Place the leaf on the middle of the microscope slide, drop 1-3 drops of water on top of the leaf, and place the cover slip over the leaf. When placing the cover slip place it at an angle to avoid too much air being trapped. Remove any excess water gently with a Kimwipe.

Using the Microscope:

1. Place the slide on the stage and rotate the objective to the lowest power (smallest lens). Starting at the lowest power, view the plant cells. When focusing the microscope raise the stage up toward the objective, then looking through the eyepiece lower the stage until the image is focused.

2. Move the objective to the next higher power by rotating the nosepiece; do not grab the objectives to rotate. If you have not moved the stage, the image should be near in focus.

3. If the microscopes are already set up, just view the plant cells under high and low power. Only adjust the focus using the fine focus knob. If the image is lost, contact the instructor

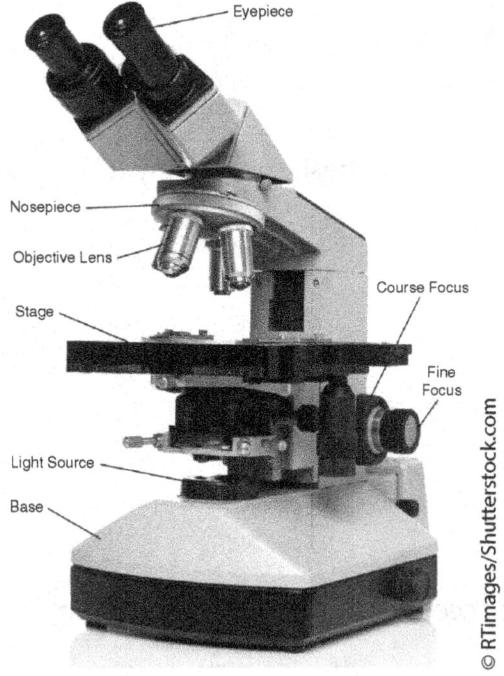

Figure 7.3   Parts of the microscope.

4. In the space in the next page draw the image that you see and label the chloroplast and cell wall in your drawing. Add an explanatory title or caption to each image.

5. Observe bacteria cells under the microscope and note the differences between the plant cells and the bacteria cells.

6. Besides color, describe two observed differences in cell structure or cell components between bacteria cells and plant cells.

7. In the space below, draw the plant cells in low power and high power settings (***label chloroplast and cell wall, add title/caption to each drawing***):

Part C: Separating Photosynthetic Pigments

In this experiment, the photosynthetic pigments from the spinach leaves will be extracted and separated using the technique of paper chromatography. Chromatography is a method used to separate mixtures and identify their components. There are various types of chromatography (column, paper, thin-layer, gas), but in all cases the separation is achieved by distribution of components between a fixed or stationary phase and a moving or mobile phase. In paper chromatography, the components of a mixture are separable into discrete zones on a sheet of chromatography paper.

Chromatography works by taking advantage of both capillary action (flow of liquids through a porous media) and polarity differences between molecules. Polarity refers to the distribution of electrons in a molecule. A polar molecule has one part that is slightly negative as another part of the molecule is slightly positive, whereas nonpolar molecules have equal distribution of electrons and no partial charges exist. Using chromatography paper as a polar stationary support (cellulose and water) and a solvent as a mobile nonpolar phase (9:1, acetone:hexane solution), the photosynthetic pigments will be carried along the paper as the nonpolar solvent moves up the paper through capillary action.

The separation of the pigments will be dependent on their affinity for the polar (cellulose and water) versus the nonpolar solvent. Those pigments that are highly polar will travel the least distance, while those that are nonpolar will travel the greatest distance along the paper.

The various pigments have different numbers of polar groups attached to them: **chlorophyll a** has 5 polar groups; **chlorophyll b** has 6 polar groups; **carotene** has 0 polar groups; and **xanthophyll** has 2 polar groups.

Using the information presented above, make a hypothesis about how the pigments will separate when you carryout paper chromatography of a spinach leaf.

H:

Chromatography Procedure

1. Each student will get two pieces of chromatography paper.

2. Draw a PENCIL line 2.0 cm from the bottom of each paper strip. BE SURE TO KEEP THE STRIP ON A PIECE OF PAPER AND DO NOT TOUCH THE PAPER WITH YOUR FINGERS. Place your initials in pencil at the top of the chromatography paper.

3. Place a spinach leaf on top of the chromatography paper and using a glass rod grind the leaf into the paper along the pencil line (be certain not to tear the paper). You need to repeat this step with a "fresh" section of the leaf 3 times or more. You should now have a nice "grass" stain along the line on the chromatography paper. The darker your line, the better your results. Repeat this procedure for your second piece of chromatography paper.

4. Carefully take the chromatography papers to the chromatography jars. Each jar will have 2 cm of solvent on the bottom of the vial. The instructor will place the paper in the chromatography jars and close the jars. **The pigment line (grass stain) cannot be immersed in the solvent.** THE SOLVENT IS HIGHLY FLAMABLE AND TOXIC—DO NOT INHALE THE FUMES FROM THE SOLVENT!!

5. Allow the chromatography to run for 10-15 minutes or when the solvent is within 2 cm of the top of the paper. Carefully remove the paper and check for separation of pigments.

6. Attach each of your chromatography papers to a separate piece of paper or lab page to turn in with the lab report.

7. Identify then label the pigment bands by their colors and relative positions on the chromatogram paper for each run. Chlorophyll b is olive-green; chlorophyll a is blue-green. There should be two bands of yellow carotenoids; a "fast" band (carotene) and a "slow" band (xanthophylls). **Outline and label each pigment**.

8. Measure the distance from the point of application to the pigment band and **write** this next to each pigment name.

9. Calculate and record the retention factor or Rf values for each pigment using the equation below and input the results on Table 7.2 (**show all your work**). The greater the Rf value the more nonpolar the pigment.

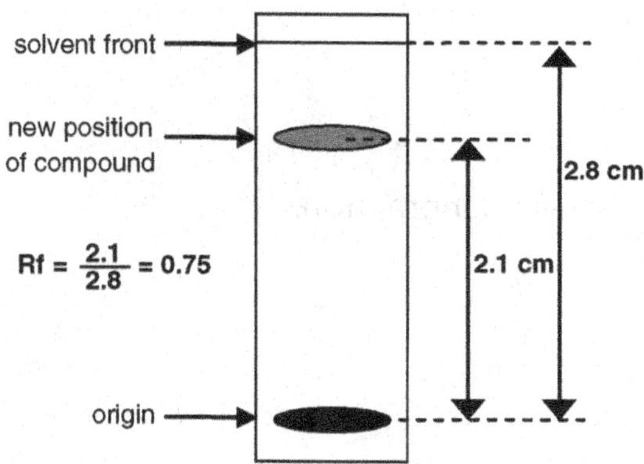

$$Rf_{(unitless)} = \frac{\text{the distance traveled by the pigment (cm)}}{\text{the distance traveled by the solvent (cm)}}$$

Results:

Attach your chromatography sheets to the back, clearly label each of the pigments and distance traveled for each. Show your work.

Table 7.2   Results for Pigment Separation Using Paper Chromatography

Pigment	Rf value, Run 1	Rf value, Run2

Answer the following questions:

1. Did you accept or reject your hypothesis? Why?

2. What benefits do the plants get by having several pigments (why are there several pigments in green leaves)?

3. During the fall leaves change colors; how does this relate to photosynthetic pigments?

4. Pose a hypothesis to explain this phenomenon.

5. How would you test the hypothesis developed above?

Place chromatography paper here. Outline each pigment, label the pigment, and write the distance travels (in cm) for each pigment. Show how you calculate Rf for each pigment.

Risk Assessment

Objectives

1. Calculate simple probability.
2. Identify and discuss voluntary and involuntary risks.
3. Discuss the difference between real and perceived risk.
4. Identify dose response curves and positions on a curve.

Materials Needed

Six-sided die
Coins
Colored pencils

Select and write one
number in the box below:
1, 2, 3, or 4.

Introduction

Risk can be described as the possibility of suffering harm or loss from a hazard resulting in injury, disease, death, economic loss, or environmental damage. A **hazard** is the source of the risk. To have a risk you must be exposed to that hazard. Some examples of cultural hazards are unsafe working conditions, smoking, poor diet, and drinking. Chemical hazards refer to harmful chemicals in the air, water, soil, or food. Biological hazards are those from bacteria, viruses, pollen or other allergens as well as poisonous animals (e.g., snakes, bees). Lastly, physical hazards include fire, tornadoes, hurricanes, floods, volcanic eruptions, and earthquakes.

Risk assessment is the scientific process of determining how much harm a particular hazard poses to human or ecological health. Risk is usually expressed mathematically as probability or the likelihood of an event occurring. For example, the lifetime probability of developing lung cancer from smoking a package of cigarettes a day is 1 in 250. This means that 1 out of every 250 people that smoke a package a cigarette throughout their lifetime (usually 70 years) will develop lung cancer. This can also be expressed as a number ranging from 0 to 1. An impossible event or no risk has the probability of 0 and the maximum risk or inevitable event has the probability of 1, and every other possibility is between 0 and 1. A six-sided die with sides numbered from 1 to 6 can be used to examine probability. The probability of rolling a single die and getting a number 2 is 1 chance out of 6. The probability is 1 divided by 6, or 0.1667, this can be expressed as 16.67%. If you were to roll a die 100 times, you would expect to get the number 2 between 16 and 17 times.

The types of hazards we are exposed to depend on our activities, habits, diet, genetics, and many other factors. Some of these hazards create **voluntary risks** which are risks we voluntarily choose to take while **involuntary risks** are risks we do not voluntarily choose to take. People do not like to be forced to face a risk such as trace chemicals in tap water, but they will voluntarily assume risk like eating a high fat diet or smoking. Risks we choose to take usually don't seem as scary as those imposed upon us by someone else when, statistically speaking, the opposite may be true. **Risk perception** involves people's judgment about the type and severity of a risk. Emotions play a large role in public perception of risk. When people become aware of a threat, they want to maintain control and protect their family and property. **Perceived risk** is that risk that a person believes exists regardless of whether or not any real risk exists. A person's belief of how severe a risk might be is also included as part of perceived risk. An example of perceived risk is that most people believe that flying or driving a car is relatively safe and bungee jumping is very dangerous. If bungee jumping was very dangerous, we would see large amount of people being injured and/or dying each year from the activity. The fact is very few bungee jumping accidents occur. **Real risk** is the actual statistical likelihood of an adverse effect happening. When evaluating risk for ourselves it is important to examine how we acquire our perceptions and whether or not they accurately represent an activity's real risk.

Chemical risk assessments are used to determine how and when a chemical or pollutant poses an unacceptable risk. All substances can be toxic at some level. Paracelsus (Phillip von Hohenheim 1493–1541), sometimes called the "father" of toxicology, wrote that *all things are poisonous and nothing is without poison, only the dose permits something not to be poisonous.* That is to say, everything is toxic at some dose, even things we think of as safe. For example water and sugar, if taken in large enough dose cause a fatal effect. A *toxin* is a poisonous substance produced by a living organism or cell. A *toxicant* is a chemical compound that has an effect on living organisms; many of these compounds are released into the environment by human activity. The effect or how poisonous the compound is depends on the concentration or dose of the compound, which is a measure of the compound's **toxicity**. Toxicity tests are laboratory tests that examine the response of different doses of a specific chemical or toxicant and the resulting effect. The results of these tests are called a **dose-response curve** (Figure 9.1). The dose-response curve graphically represents how the severity of an effect changes with increasing doses. There are three general stages in severity illustrated by a non-cancerous dose-response curve (Figure 9.1): stage 1 = no effect range, stage 2 = range of increasing effect with increasing dose and stage 3 = maximum effect range. Toxicity tests are used in combination with exposure assessments to prepare a risk assessment which describes what is likely to happen to a person in a real world situation. Governmental regulations for air pollution, water quality, food quality, and work place environments are based on risk assessments.

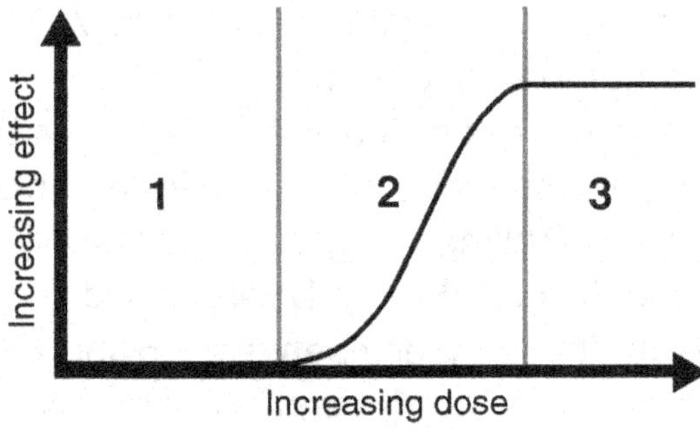

Figure 9.1   Representation of the dose-response relationship.

1. No-effect range
2. Range of increasing effect with increasing dose
3. Maximum effect range

Part B: Determining Probability

19. In groups identify three *voluntary* risks and three involuntary risks you face daily.

Involuntary Risk	Voluntary Risk

Health Risk Assessment

20. Use the Table 9.4 to assign yourself risk points for developing diabetes disease in the next 10 years. If you do not meet the criteria, insert a "0."

Table 9.4   Diabetes Risk Assessment (If you do not meet the criteria, insert a "0.")

Diabetes Risk Factor	Category	Impact	Your Risk
Family history	Yes	+2	
African American, Latinos, Native Americans, Asians and Pacific Islanders background		+2	
Weight factor, Women	BMI* 23-24.9	+2	
	BMI* 25-26.9	+3	
	BMI* >27	+4	
Weight factor, Men	BMI* 23-24.9	+2	
	BMI* 25-29.9	+3	
	BMI* >30	+4	
Waist size	Women > 35 inches	+3	
	Men > 40 inches	+4	
Active smoking	1-14/day	+1	
	> 14/day	+2	
Physical activity	30 min. /day or > 3 hrs/week	-2	
Cereal fiber/Whole grains	> 3 svg/day	-2	
Refined grains	> 3 svg/day	+1	
Mono and polyunsaturated fats (liquid vegetable oil, oil-base salad dressing)	> 4 svg/week	-1	
Alcohol Intake (no binging, ~1/day)	2-3 svg/week	-2	

BMI = Weight (pounds)/Height (inches^2) * 703

21. Your total risk points for Diabetes ____________________

22. Divide the total risk points by 4 (the estimated population average score) to get your score. Using the scoring key (Table 9.6) what is your level of risk for diabetes?

23. Use the Table 9.5 to assign yourself risk points for developing heart disease in the next 10 years. If you do not meet the criteria, insert a "0."

Table 9.5   Heart Disease Risk Assessment

Heart Disease Risk Factor	**Category**	**Impact**	**Your Risk**
Sex	Male	+2	
Family history (Immediate)	Yes	+2	
Weight factor, Women (< 60 yrs.old)	BMI* 25-28.9	+2	
	BMI* > 29	+3	
Weight factor, Men (< 65 yrs.old)	BMI* 25-29.9	+2	
	BMI* > 30	+3	
Waist size	Women > 35 inches	+2	
	Men > 40 inches	+2	
Active smoking	1-14/day	+1	
	15-25/day	+2	
	>25/day	+3	
Past smoking (Cessation duration)	< 10 years	+1	
Passive smoker	Regularly exposed	+1	
Diabetes or high blood sugar	Occurrence	+2	
Hypertension or high blood pressure	Occurrence	+2	
Total cholesterol*	160-199	+1	
	200-239	+2	
	240-279	+3	
	>280	+4	
Physical activity	> 3 hrs/week	-2	
Fish intake	> 2 svg/week	-2	

If unknown, assume lowest value or use family history.

Heart Disease Risk Factor	Category	Impact	Your Risk
Fruit/Vegetable intake	> 5 svg/week	-2	
Cereal fiber/Whole grains	> 3 svg/day	-2	
Nut intake	> 3 svg/day	-1	
Saturated fat (butter, lard, red meat, cheese, whole milk)	> 3 svg/week	+1	
Transunsaturated fats (margarine, vegetable shortening, deep fried food)	> 4 svg/week	+1	
Mono and polyunsaturated fats (liquid vegetable oil, oil-base salad dressing)	> 4 svg/week	-1	
Alcohol intake (no binging, <1/day)	2-4 svg/week	-2	
Multivitamin / B complex supplement	> 4 times/week	-1	
Vitamin E	> 4 times/week	-1	

BMI = Weight (pounds)/Height (inches^2) * 703

24. Total Risk points for Heart Disease ____________________

25. Divide the total risk points by 3 (the estimated population average score) to get your score. Using the scoring key (**Table 9.6**) what is your **level of risk** for heart disease?

Table 9.6   Scoring Key for Level of Risk for Diabetes and Heart Disease

Risk Score	Level of risk
<0	Very much below average risk
0 or <0.5	Much below average risk
0.5 to < 0.9	Below average risk
0.9 to <1.1	About average risk
1.1 to < 2.0	Above average risk
2.0 to < 5.0	Much above average risk
5.0 or more	Very much above average risk

Adapted from *Your Disease Risk*, Harvard Report on Cancer Prevention, Volume IV: Harvard Cancer Risk Index, Cancer Causes and Control, Volume 11: 477–488, 2000.

26. Regardless of your score, list four ways you can reduce your risk for diabetes and heart disease.

For more information on disease and cancer risk and how to reduce your risk go to http://www.yourdiseaserisk.wustl.edu/index.htm

Composting Laboratory Exercise

Objectives

1. Understand the process of composting and how it is important in nutrient cycling.
2. Understand how the carbon to nitrogen ratio affects composting rates.
3. Learn how to measure various characteristics related to composting to determine the activity of the compost.

Materials Needed

Quart size or larger container
scraps newspaper or similar
1/2 to 1 cup soil
Aluminum drying pans
Compost material (cut into less than 1/2" size)

Introduction

Composting is a way to add valuable nutrients and humus back to soil. It can be done with any organic material, but most commonly nonanimal organic waste including kitchen scraps and leaves and yard waste is used for composting. Compost is made when organic material is combined in proper ratios to accelerate the breakdown of the organic materials. During the process of composting the organic material is broken down by microorganisms under aerobic or oxygen-rich environment. The decomposition process breaks apart the organic material to produce carbon dioxide (CO_2), water vapor (H_2O), and heat (Figure 10.1). The product is a stable, fine grain, material called humus. Humus is fine organic material that is dark brown or black, has an earthy smell and is rich in plant nutrients. There are different types of **aerobic bacteria** that are active in composting. The populations of bacteria vary by the temperature of the compost. At temperatures less than 70° F, **psychrophilic** bacteria are found in compost piles. They are most active at 55° F and give off a small amount of heat in comparison to other bacteria. **Mesophilic** bacteria rapidly decompose organic material; their working temperatures range between 70° and 100° F. When temperature exceeds 100° F, mesophilic bacteria start to die off and thermophilic bacteria become abundant. **Thermophilic** bacteria thrive at temperatures from 113° F to 160° F. Once the compost pile reaches thermophilic temperatures, new material must be constantly fed to the pile and turned at strategic times. These high temperature conditions usually last three to five days and then as the temperature declines, mesophilic bacteria become dominant once again. Effective decomposition of compost occurs under mesophilic conditions but optimal decomposition takes place under thermophilic conditions.

Besides temperature changes in the compost pile, oxygen and moisture conditions can make bacteria die or become inactive. When **oxygen levels** decrease to less than 5%, anaerobic bacteria take over and decomposition slows greatly. **Anaerobic bacteria** produce organic acids and hydrogen sulfide which release offensive odors, an indication of anaerobic conditions. Microorganisms also need adequate amounts of moisture to survive. **Moisture** or water provides a way for the nutrients and organic material to be accessed by the microbes. Ideally, compost piles should contain between

40 and 60% moisture by weight. At lower levels (<30%), microbial activity and degradation rates are slowed. At moisture conditions >65%, decomposition rates and microbial activity also slow with the potential formation of pockets of anaerobic bacteria as air spaces fill with water.

Ohio passed a **law** in 1995 to ban the disposal of yard waste in landfills. **Yard waste** is solid organic waste that includes leaves, grass clippings, brush, garden waste, holiday trees, and pruning from trees or shrubs. Burying yard waste in landfills is not cost effective way to use valuable landfill space. Yard waste buried in a landfill under trash takes years to decompose and the important nutrients it contains are lost. Composting yard waste converts raw organic matter into a decomposed product that releases nutrients into soil in a faster, steadier way than raw organic matter. The nutrients released by compost improve quality of the soil.

For effective composting, a compost bin must include both "**brown**" or carbon (C) and **"green"** or nitrogen (N) organic waste. The brown are dried plant material such as straw, dried leaves, and woody debris. These materials are primarily made up of long chains of sugar molecules linked together (cellulose). These items are a source of energy for the compost microbes. "Greens" are fresh (and often green in color) material such as vegetable scraps, fruit scraps, green leaves, coffee grounds, and tea bags (Table 10.0). These items contain more nitrogen in them as compared to the "browns." Nitrogen is a critical component of amino acids which are needed to make proteins. Proteins are needed in many metabolic activities in all living organisms. Too little nitrogen and your pile will decay into compost a lot more slowly. A compost bin that doesn't heat up after 24 hours may need more "green" material. Therefore, a good mix of browns and green will provide balanced energy and nutrients for the microbes.

For more information see these websites:

http://www.epa.gov/compost/basic.htm
http://web.extension.illinois.edu/homecompost/science.cfm
http://whatcom.wsu.edu/ag/compost/fundamentals/
http://compost.css.cornell.edu/

Figure 10.1   The components of composting include an input of raw organic material (nitrogen and carbon) and water. Microorganisms breakdown the organic material to release heat, carbon dioxide, and water vapor. The product is finished compost or humus.

Hypothesize on the rate of decomposition of your compost based on C/N ratio and the amount of material used. In your hypothesis indicate how you will determine the rate of decomposition (e.g., determined by how much of the material can be identified or looks like original material).

H:

Composting at Home

Methods: How to start building your compost:

1. Start with a plastic container and lid that is quart size or larger. If the selected container doesn't have a lid or seal, plastic wrap/a plastic bag and a rubber band can be used to cover the container. Be sure to rinse the container out completely. This is especially important if using containers that once held dairy or fats.

   a. Some containers that can be used include soda bottle, milk container, plastic food containers (think a take-out soup container), plastic shoe box sized container.

   b. If the container does not have a wide mouth, make a funnel out of the top of another plastic bottle or fold a piece of paper to help when inserting the composting materials. You can also, cut the top of the 2L plastic bottle and invert it as a funnel and cover.

2. Along the bottom of the container layer ½ - 1" of soil. The soil should be a dark, rich soil such as what would be found where plants are actively growing. If you do not have soil available, please inform your instructor so some can be provided.

3. Add compostable food scraps to your compost experiment and record the amount and ratios outlined in Table 10.1. Compostable material are primarily vegetable and fruit scraps but can include dried leaves or grass to achieve the proper brown: green ratio (C:N ratio). If your lifestyle doesn't include many compostable material, consider what is left on your plate after you eat. Wilted lettuce from your cheeseburger, or the tomatoes that fall off from your taco can all be added to your compost. Ask family and roommates for help too.

   a. Things that *should not* be included in compost are dairy and fats. If scraps intended for compost have butter, dairy, dressing, or sauces which contain those components, rinse the scraps before adding them. Additionally, meat products should not be added to your compost.

   b. The size of the food scraps matters. You will not want to add large pieces since we are creating a mini-compost bin. The smaller the size of the scrap the better. For example, a banana peel will need to either be cut or torn into half inch sized or smaller pieces.

   c. Certain materials such as egg shells should be added sparingly. Limit to half an egg shell and make the pieces as small as possible. The high calcium content of egg shells can increase the pH of the compost to a level which would render it unusable on plants.

4. Add food scraps for 4 weeks or until your container reaches ½ to ¾ capacity. If you are using a wide mouth container, you will want to add a layer of shredded newspaper (or other lightweight paper) to prevent flies from becoming a nuisance. Complete Table 10.2 to track the amount and type of compost added to your mini-compost system.

5. After food scraps have been added to reach about ¾ full, you will need to allow the microbes to work breaking down the material.

6. <u>Where to keep your compost:</u> Composting microbes prefer ambient temperature for optimal composting. Do not place your compost in direct light or near the heater/AC vent. Place mini-compost bin in darker area that has a stable temperature.

7. Remember to record what you are adding to your compost and approximately how much of each. Should your compost decompose slowly, it might be due to an unbalanced ratio of green to brown.

Table 10.1   Log of composting material added to the mini-compost bin

Date Added	Material	C:N Ratio	Estimated amount

Data Collection/Observations:

Observations should be made weekly! Remember to date your information.

1. Make visual observations on:
    - The height of your compostable materials. Is there any change in height? Record it on your bottle if necessary.
    - Keep track of any condensation or liquid you see forming in the container
    - Amount and type of un-decomposed organic material (OM)
    - Color & texture – getting darker? Slimy? Becoming granular?
    - Granular size & composition – are things breaking down? Which is breaking down faster, the brown or green material?
    - Any other observations, (plants growing, odor, water content high or low, etc.)

    Observations can be kept in a table (see example in Table 10.3) or simply as a dated list of observations. Use a method that allows you to provide detailed and thorough information about your observations. Taking photos of your experiment throughout the process can be beneficial in helping you describe the composting process and the rate of weekly decomposition. You can also use photos as a reminder of what materials were added to the compost.

Table 10.2   An example of a mini-compost observations

Date: 1/25/2020	
Week 5: no additions to compost.	
Height	Initial height was 13 cm, this week 12.5 cm
Compositional changes	Banana peels are brown on both sides, the orange peel appears withered – no color change yet, Diced tomato appears to be almost gone, only the peel is truly visible. Brown material exhibits no change although the green material is showing decomposition
Moisture Content	Condensation is forming in the bottle, definitely an increase in moisture from last week.
Color/Texture	No large color changes, some decay is apparent. The tomato is breaking down quickly.
Granular size	Organic content is breaking down, no obvious granular development yet.
Odor	No odor at this time
Other:	Air pockets visible through sides of bottle, no need for aeration at this time

2. After about 8 weeks, you will be instructed to bring in your mini-compost bin for further analysis.

3. Determine percent moisture and pH of compost.

   a. Percent Moisture:

      i. Weigh the drying pan and record the weight: _______________________g

      ii. Collect a compost sample from you bin from the middle of the container and about 1" in depth and place in pre-weighted drying pan.

      iii. Weigh the moist compost in the drying pan: _______________________g

      iv. Place pan and compost in the drying over for 24-48 hours at 100-105°F.

      v. Remove the drying pan and compost and allow to cool to room temperature.

      vi. Weigh and record the total weight of pan and compost: _______________________g

      vii. Determine the weight of moist compost and dry compost by substacting the weight of the drying pan.

      viii. Calculate the % Moisture using Equation 1.

$$\% \text{ moisture} = \frac{(\text{wet weight} - \text{dry weight})}{(\text{wet weight})} \times 100 \qquad \text{Eq. 1}$$

   b. Compost pH

      i. Press pH paper into moist soil and record pH by comparing it to pH chart.

Results:

A lab report or research report is composed of various section to allow the reader to follow the experimental process and what was concluded.

Each student will answer questions via Blackboard (or other online format) about the composting experiment including questions about each section of a lab report. For example, questions on

- Introduction: Background information on composting.
- Materials & Methods: Information on how the experiment was done including type of compost material, environment it was kept in, any problems during the experiment.
- Results: Your observations, percent moisture and pH, decomposition rate.
- Discussion: Was the hypothesis accepted or rejected and why? What influenced how fast or slow the material composted? What could you do to improve the experiment?

Methods

B. Percent (%) Moisture—Over the same three week period, you will measure and record the percent moisture of the compost at one location in the bin. The compost percent moisture will be measured following the steps listed below.

1. Collect a compost sample from the top of the compost/soil off the surface about 1 inch, then pull a sample from the bin and place in an aluminum drying pan.

2. Bring the drying pans filled with compost back to the lab.

3. Weigh and record the wet weight for each drying pan filled with compost.

4. Place the drying pan filled with compost in the lab oven.

5. Bake the drying pan filled with compost for 24-48 hours at 100-105° C.

6. Remove the drying pan filled with compost from the oven.

7. Weigh and record the dry weight for the drying pan filled with compost.

8. Remove dry compost from drying pan and place it in the container indicated by the instructor.

9. Weigh and record the weight of an empty drying pan.

10. Calculate percent moisture.

Percent moisture = (wet weight – dry weight)/wet weight x 100   **Eq. 2**

Caution! The weight referred to in the equation above is the weight of the compost only. To get the weight of the compost alone subtract the weight of the empty drying pan from the weight of the drying pan with wet compost and the weight of the drying pan with dry compost.

C. Compost pH—Press the pH paper into moist soil and record pH.

D. Record air temperature (both °F and °C) and note location of measurement.

Table 10.3   Partial Listing of "Green" and Brown" Compostable Materials (totals available on Blackboard)

"GREEN" MATERIAL	C/N Ratio
Coffee grounds and filter	25/1
Fruit peels and rinds	25/1
Garden debris, fresh	20-30/1
Grass clippings, fresh	17/1
Plants and plant cuttings	20-30/1
Pumpkins	20/1
Vegetable scraps/peelings	12/1
Tea leaves	15/1

"BROWN" MATERIAL	C/N Ratio
Corn stalks	60-75/1
Tea bags	35/1
Garden debris, dried	60/1
Grass clippings, dried	50-80/1
Hay	30/1
Leaves	50-80/1
Peanut shells, crushed	35/1
Straw	80/1
Sawdust	400-500/1
Woody chips and twigs	25/1

Table 10.4   Estimate of Overall C/N Ratio for Material Added to Compost Bin

Compost Material	Estimated Overall C/N Ratio	Total Mass Added (g)
Brown		
Green		

Part B Instructions:

In this exercise, we will use the Simpson Diversity Index to compare the amount of diversity of two communities of mollusks.

Scenario

The South Bay has an industrial effluent entering the bay at **point Z**. The effluent contains components that are heavier than water, and therefore sink to the bottom (benthic layer) of the bay. This effluent has been kept at or below permitted level but due to the heavier components of the effluent there is some concern that the effluent could have a negative impact on the benthic invertebrate communities. Using the Simpson Index, you will be comparing the amount of diversity at two sites in South Bay. **Site Z samples** come from a location where the effluent is released. **Site Y (Control) samples** come from a site in South Bay where there is no effluent from any industry released.

Molluscs or mollusks are a marine invertebrate of the phylum Mollusca that have soft unsegmented body usually enclosed in a shell. Mollusks include clams, scallops, snails, chitons, oysters, squid, and octopuses. Many mollusks live most of their live on the bottom of the bay or in the top 3-5 inches of sand or sediment. Gastropodes comprise the largest and most varied of the mollusk groupings including limpets, abalone, and snails. Depending on the species, they will consume food ranging from algae to fish, whereas bivalves, such as clams, oysters, and scallops, are suspension or filter feeders taking in large quantities of food-laden water through the mantle cavity and mouth. Diversity will be determined using mollusk because of where they reside and ease of sampling.

Develop a hypothesis (*H2*) that describes whether there is impact on the bay community due to the effluent.

H2:

1. Before depositing the samples, examine the site and decide which six squares you will sample.

    Identify sampling squares here:

    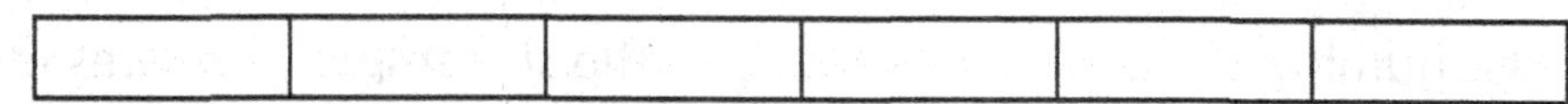

2. Select the container labeled "Y." Scatter the shells from the bag onto the site grid, this is the Y (control) sampling site. Do not move the organisms or the grid once they have been dropped.

3. Count the number of each species found in the sampling squares identified above; input the results in Table 11.8.

4. Calculate the relative abundance and relative abundance^2 for each species (Eq. 2).

Table 11.8   Community Y: South Bay Control Site, Six (6) Sample Points

Species	Number in Sample	Relative Abundance	Relative Abundance^2
Total			

5. Calculate the **Simpson Index** (Eq. 1) from the data in Table 11.8.

6. Clear off your grid and put away the organisms

7. Select the container labeled Z. Drop the shells from the bag onto the site grid; this is the Z sampling site. Do not move the organisms or the grid once they have been dropped.

8. On first inspection, does this community look more diverse, less diverse, or the same diversity as the Y sampling site?

9. Count the number of each species found in the *six* sampling squares identified previously; input the results in Table 11.9.

10. Calculate the relative abundance and relative abundance^2 for each species (Eq. 2).

Table 11.9   Community Z: South Bay Effluent Site, Six (6) Sampling Points

Species	Number in Sample	Relative Abundance	Relative Abundance^2
Total			

11.   Calculate the **Simpson Index** (Eq. 1) from the data in Table 11.9.

Part B Questions:

1.   How confident are you that your 6-square sample adequately sampled the diversity of these two communities? *Why or why not?*

2.   Which communiy had the highest species richness? Explain.

3.   Which community had the highest species evenness? Explain.

4.   Which community had the highest Simpson Index? Did this match with your first impression?

5.   Did you accept or reject your hypothesis (*H2*)? *Why?*

6. Based on your experience, would you recommend increasing, decreasing or keeping the same number (6) of squares for future biodiversity comparisons? *Why* (give support for your answer)?

Table 11.10   Results from Multiple Biodiversity Sampling

Group	Site Y Diversity Index	Site Z Diversity Index	Type of Sampling*
1			
2			
3			
4			
5			
6			

*Separated, grouped, or semi-groups (some grouped squares with some ungrouped squares)

7. Compare your results to other groups in your lab and record their observations in the Table 11.10. Did your group have similar results as other groups?

8. If you were to repeat the experiment, what type of sampling would you recommend based on the class data.

9. Develop another hypothesis (does not have to be on mollusk) to test whether the effluent is having an environmental effect.

10. Write a conclusion on the effluent's effect on the mollusk community based on the data and information gained in your diversity testing.

Generation and Remediation of Acid Mine Drainage

Objectives

1. Learn how to measure pH and the difference between and acid and a base.
2. Learn the sources of acid mine drainage.
3. Learn the process for neutralizing acids.
4. Learn the process for remediating or neutralizing acid mine drainage.

Materials Needed

Beakers
Stir rods
pH meters
Vinegar
Antacid
Baking soda
Deionized water
Tap water
Limestone
Funnels
Filter paper
Funnel rack and stands
Acid mine water

Introduction

Acids and bases (alkaline) were first described in the seventeenth century by the amateur chemist Robert Boyle. **Acids** were characterized as having a sour taste, corrosive to metals, change litmus (a dye extracted from lichens) red, and become less acidic when mixed with bases. **Bases** were described as feeling slippery, changes litmus to blue, and become less basic when mixed with acids. It wasn't until the late 1800s when scientist Svante Arrhenius hypothesized that acids were substances that dissolved in water to release hydrogen ions (H^+) into solution and bases released hydroxides ions (OH^-) into solution. This definition was further refined by Johannes Brønsted and Thomas Lowry in the 1920s. Any substance that can donate a hydrogen ion is an acid. A base is defined as any substance that can accept a hydrogen ion.

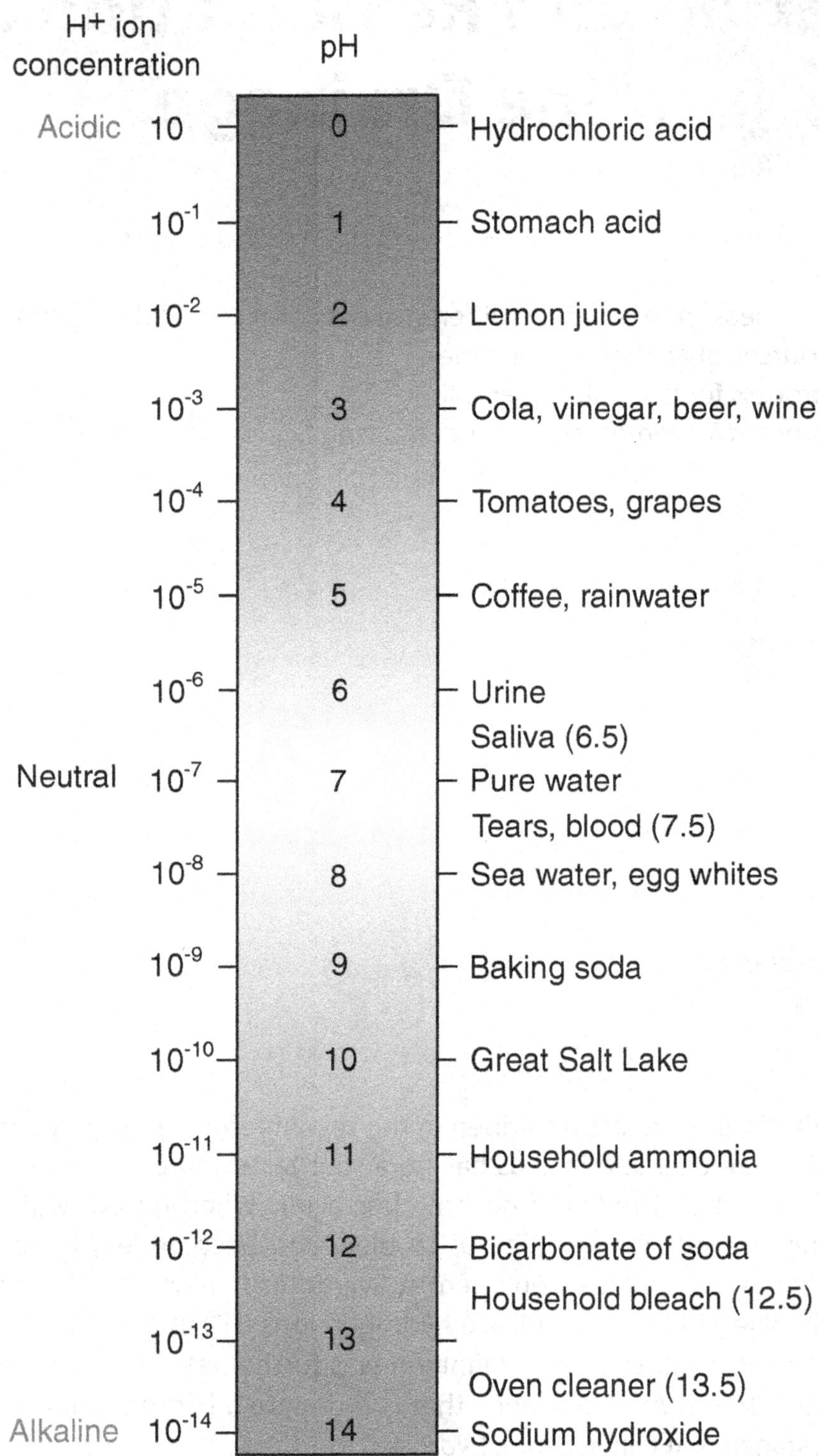

Figure 12.1   pH scale with pH noted on some common household food and products.

By measuring the concentration of hydrogen ions in a solution, it can be determined if the solution is an acid or a base. Acids increase the concentration of hydrogen ions whereas bases decrease the concentration of hydrogen ions. The **pH** of a solution is a measure of the concentration of hydrogen ions in a solution.

Water Quality

Objectives

1. Identify the characteristics of good or clean water quality for protection of aquatic life, recreation or water supply.
2. Identify the types of contaminants found in water and their effects on human and ecological health.

Materials Needed

Samples of untreated water
Test strips (5-in1, nitrate/nitrite, phosphate, and sulfate)
Colilert reagent vials
Beakers
Incubator
Ultra-violet light
Conductivity meter
Ammonia test kit
Dissolved Oxygen test kit
Water samples: Each students will provide an unregulated, untreated water sample such as a pond, stream, river or similar. To take a water sample, using tap water clean out an empty recyclable water bottle. Submerge the bottle into the water body, cap sample, invert several times, and empty the water away from sampling point. Submerge the bottle again until bottle is full, cap bottle, and refrigerate or keep cool until testing.

Introduction

Surface and ground waters have a variety of human and ecological uses. Drinking water is obtained by the cleaning and purifying of various source waters. Fishing industry and spawning sites are maintained through monitoring of a variety of chemical and biological parameters. Agriculture will use water for irrigation of crops. And many areas are used for seasonal recreation of swimming, scuba diving, boating, fishing, and beach going. All these uses have different water quality requirements that need to be met and maintained to have a sustainable system.

Some of the parameters used for monitoring and regulating water include pH, dissolved oxygen, hardness, amount of ions such as sulfates and nitrates, concentration of heavy metals, and bacteria levels.

The measure of acidity is expressed as **pH**. The range goes from 0 to 14, with 7 being neutral. A pH of less than 7 indicates acidity, whereas a pH of greater than 7 indicates a base. Since pH can be affected by chemicals in the water, pH is an important indicator of water that is changing chemically. Pollution can change water's pH, which in turn can harm animals and plants living in the water. For instance, water coming out of an abandoned coal mine can have a pH of 3, which is very acidic and would affect any fish trying to live in it.

You can't tell by looking at water that there is oxygen in it. But, if you look at a closed bottle of a soft drink, you don't see the carbon dioxide dissolved in that—until you shake it up and open the top. The **dissolved oxygen** in lakes, rivers, and oceans is crucial for the organisms and creatures living in it. As the amount of dissolved oxygen drops below normal levels in water bodies, the water quality is harmed and creatures begin to die off. Rapidly moving water, such as in a mountain stream or large river, tends to contain a lot of dissolved oxygen, while stagnant water contains little. Bacteria in water can consume oxygen as organic matter decays. Thus, excess organic material in our lakes and rivers can cause an oxygen-deficient situation to occur. Aquatic life can have a hard time in stagnant water that has a lot of rotting, organic material in it, especially in summer, when dissolved-oxygen levels are at a seasonal low.

The amount of dissolved calcium and magnesium in water determines its "**hardness.**" Water hardness varies throughout the United States. If you live in an area where the water is relatively "hard," then you may notice that it is difficult to get lather up when washing your hands or clothes. And, industries in your area might have to spend money to soften their water, as hard water can damage equipment. **Alkalinity** refers to the capability of water to neutralize acid. It acts like a buffer in freshwater systems absorbing the excess H^+ ions and protecting the water body from fluctuations in pH. The presence of calcium carbonate ($CaCO_3$) and magnesium carbonate ($MgCO_3$) contribute to this buffering system. When most of the alkalinity is due to the presence of $CaCO_3$, then hardness in $CaCO_3$ is equal to alkalinity. Hard water contains high levels of metal carbonates and is high in alkalinity whereas soft water usually has low alkalinity.

Sulfate and **Nitrate** are other chemicals that are commonly monitored in water. High concentrations of sulfate in drinking water can cause diarrhea and dehydration. At high levels (10 mg/L) nitrate-N (NO_3-N) is a health hazard to infants and pregnant women. The high nitrate level causes a condition in infants called **methemoglobinemia** or "**blue-baby**" syndrome in which there is a reduction of oxygen carrying capacity of the blood. In ecological systems excess of nitrates and other limiting nutrients (**phosphates**) can cause eutrophication. The increase in nutrients stimulates excessive plant growth (algae and nuisance weedy plants). This enhanced plant growth, often called an algal bloom, reduces dissolved oxygen in the water when dead plant material decomposes and can cause other organisms to die. **Phosphates** occur naturally in phosphorus rocks and are needed by all living cells promoting the growth of all plant life. The major cause of excess phosphates, and eutrophication, is run-off from agricultural areas, animal waste, industrial waste, and areas where phosphate fertilizers have been applied.

Ammonia (NH_3) is another source of nitrogen. Even at low levels (<0.05 mg/L) ammonia can irritate fish gills. At higher levels (>0.1 mg/L) fish can experience damage to eyes, gills, and skin. Prolonged exposure or exposure to toxic levels of ammonia can cause loss of appetite, gasping at the water surface, sluggishness, and eventually death. While ammonia is not directly an issue for drinking water safety to humans, indirectly it is because of interference in disinfection for control of disease causing organisms.

Eutrophication occurs when excess nutrients, such as phosphates and nitrogen, enter an aquatic ecosystem. This can be a natural process in lakes as they age through geological time. Human activities have increased the amount of nutrients entering both fresh and marine water systems resulting in **cultural eutrophication**. The sources of nutrients are sewage and septic systems, run off from agriculture and golf courses, animal waste and other human-related activities. The excess nutrients cause algae and some plants to grow rapidly resulting in algae blooms. As the algae dies, it fall to the sediment and is decomposed by microorganisms. The decomposition process

Parameter	Warmwater Habitat	Coldwater Habitat	Exceptional Warmwater Habitat	Recreation	Public	Agriculture Use
Chloride (Cl-)/ Free Chlorine	600 mg/L max 150 mg/L 30-d Avg.	600 mg/L max 150 mg/L 30-d Avg.	600 mg/L max 150 mg/L 30-d Avg.	—	250 mg/L	100 mg/L
Total Chlorine	0.2 mg/L Kills most plankton 0.25 mg/L Only tolerant fish survive 0.37 mg/L Maximum tolerated by any fish species			—	—	—
Alkalinity	>20 mg/L, 20-200 mg/L are typical of fresh water. A total alkalinity level of 100-200 mg/L will stabilize the pH level in a stream. Levels below 10 mg/L indicate that the system is poorly buffered (LRW).			—	—	—
Phosphorus	Total phosphorus as P shall be limited to the extent necessary to prevent nuisance growths of algae, weeds, and slimes that result in a violation of the water quality criteria. Total phosphate should *not exceed* 0.05 mg/L (as phosphorus) in a stream at a point where it enters a lake or reservoir. Total Phosphate should *not exceed* 0.1 mg/L in streams that do not discharge directly into lakes or reservoirs. Levels between 0.03 and 0.1 mg/L can trigger an algae bloom.			Inputs may not exceed 1mg/L Total P. Phosphate levels greater than 1.0 mg/L may interfere with coagulation in water treatment plants.		
E. Coli	May to Oct.: Not more than 1 positive out of 5 samples within a *30 day* period. Nov. to April: Not more than 2 positive out of 5 samples within a *30 day* period.	May to Oct.: Not more than 1 positive out of 5 samples within a *30 day* period. Nov. to April: Not more than 2 positive out of 5 samples within a *30 day* period.	May to Oct.: Not more than 1 positive out of 5 samples within a *30 day* period. Nov. to April: Not more than 2 positive out of 5 samples within a *30 day* period.	Not more than 1 positive out of 5 samples within a *30 day* period.	—	—

— Indicates no levels set for this parameter for this water classification

Please note: *Majority* of these parameters need to be met to accept the classification of water use. If one or more of the parameters are exceeded, then consideration would have to be taken as to the importance of the parameter and whether the parameter that is exceeded regularly or is a single event. Further testing would need to be done to determine if that water will meet the water classification. Therefore, can accept the hypothesis for water classification as "conditional" if one parameter did not meet specifications, then state that *further testing would need to be done to determine if that parameter will meet the level needed by the water classification.*

Consider the following questions and statements when observing and describing your results and incorporate your answers into your final report. Discussions report must be 2-3 full pages, 1.5 spaced, 10-12 size font with 1" margins.

- Did you accept or reject your first hypothesis (H_1)? Why?
- Did you accept or reject your second hypothesis (H_2)? Why?
- How does the colors and sources of each of your water samples compare?
- How did your chemical and biological test compare to Table 13.3 for the proposed use of your water samples?
- What were the sources your group's samples? How did that affect your test results? Was it expected or unexpected?
- Which parameter would be of greatest concern for each water sample? Why?
- How do the levels of nutrients compare to other group's water or between your two water samples?
  - A. How will the level of nutrients affect plants and other organisms in your water system?
  - B. If the level of nutrients were high/low, what kind of visible effect could you observe? What other changes might be expected to occur?
- If any of your group's samples tested positive for Coliform or E. coli bacteria, explain how the water sample could have become contaminated or the source of the waste.
  - A. How could the microorganisms from your water samples come into contact with the population? What is the route of exposure?
  - B. What types of illnesses are associated with each type of microorganism?
  - C. Would a boil alert help to reduce these bacteria from being present? How could the effectiveness of this type of precaution be tested?

REPORT DUE _______________________________________

References:

- State of Ohio Water Quality Standards. 2017. Ohio Environmental Protection Agency, Division of Surface Water, Standards & Technical Support Section, Chapter 3745-1 of the Administrative Code. Available from: http://epa.ohio.gov/dsw/rules/3745_1.aspx.
- Division of Surface Water, Ohio EPA. 2017. Water Quality Standards Program Available from: http://www.epa.state.oh.us/dsw/
- USGS. 2016. The USGS Water Science School https://water.usgs.gov/edu/waterquality.html
- US EPA. 2017. Freshwater Quality. Available from: https://www.epa.gov/salish-sea/freshwater-quality

Aquifers and Groundwater Contamination

Objectives

1. Learn how porosity effect permeability and influence groundwater recharge.
2. Learn and describe confined and unconfined aquifers.
3. Recognize how groundwater is polluted and steps that can be taken to remediate polluted groundwater.
4. Learn how to reduce groundwater over use and groundwater pollution.
5. Understand what groundwater mining is and how it affects water availability.

Materials Needed

Groundwater model
Liquid dyes
Cylinders
Mesh
Rubber bands
3 Different gravels
Beakers
Water

Introduction

The hydrologic cycle describes the storage and movement of water above, on, and in the Earth's surface. Water **condenses** in the atmosphere forming clouds. When the water droplets in the cloud gain enough mass the water falls as **precipitation**. Precipitation (rain fall, snow, etc.) falls to the ground where it can **evaporate**, be taken up by plants or animals, **runoff** into surface water or **infiltrate** through the pores and cracks in soil and rock to the groundwater. Groundwater is the water that is stored beneath the earth's surface. Groundwater can be found at very shallow depths (close to the earth's surface) or very deep. Groundwater that is very deep could be thousands of years old once the water reaches the deep aquifer, whereas shallow groundwater may only be hours old.

Groundwater is found in the cracks and holes in moderately or highly permeable rocks called **aquifers**. An aquifer can be composed of layers of sand, gravel, sandstone, limestone, lava flows, or even fractured granite. **Permeability** is the measure of how pores and fractures in the rocks or soil transmit water and other fluids through the pores*. As water permeates through the soil

---

* **Porosity** is the amount of empty space in a rock or other earth substance; this empty space is known as pore space. Porosity is how much water a substance can hold. Porosity is usually stated as a percentage of the material's total volume.

$ Water flows between the spaces in the ground. If the spaces are close together such as in clay-based soils, the water will tend to cling to the material and not pass through it easily or quickly. If the spaces are large, such as in gravel, the water passes through quickly.

it replenishes or **recharges** the groundwater. The recharge rate varies depending on the surface conditions and the amount of land area available to allow water through. Human activities such as development, paving, and logging increase the amount of surface runoff and decrease the amount of recharge.

An **unconfined aquifer** is an aquifer that is open to receive water from the surface and whose water table fluctuates up and down depending on the recharge rates. The layer immediately above an unconfined aquifer where pore spaces are filled with both air and water is called the **unsaturated zone** or **vadose zone**. The **water table** is the top of the **saturated zone;** this layer has all the pore spaces filled with water (Figure 14.1). The level of the water table and hence the thickness of the unsaturated zone is dependent on the recharge rates and groundwater removal rates.

A **confined aquifer** is an aquifer that is saturated with water and usually deeper than unconfined aquifers. Layers of impermeable material, such as clay, are both above and below the aquifer impeding the movement of water into and out of the aquifer. Because of the impermeable layers, water in these aquifers is under pressure. When the aquifer is penetrated by a well the water will rise above the top of the aquifer. This groundwater that flows up to the surface without any pumping is an **artesian well**.

Groundwater uses

Groundwater has been used for years as a water source for human consumption, agricultural irrigation, livestock, aquaculture, industrial uses, mining, and thermoelectric power generation. There is 30 times more groundwater than all the fresh-water from lakes, although only a fraction of the groundwater can be practically used, much of the groundwater is much too deep to access (USGS 1999). **Groundwater mining** is the removal of water from aquifers for human use resulting in the lowering of the water table. This type of mining is different from mining for copper, gold, or other minerals. Groundwater, in most cases, is a renewable resource whereas minerals are nonrenewable. Approximately 26% of our nation's drinking water comes from groundwater and another 68% is used for irrigation. In some rural areas, groundwater wells are the only source of drinking water. The mining of groundwater can lead to overdraft. Groundwater **overdraft** occurs when the amount of removal exceeds the amount of groundwater recharge causing shallow wells to be unusable. Overdraft can also lead to groundwater subsidence or saltwater intrusion. **Subsidence** occurs when portions of an aquifer that no longer contains water collapses. When the subsidence occurs it lowers the level of the land surface above. For example, the city of San Jose, California lowered 13 feet in the mid-1900s due to overdraft of groundwater and subsidence. **Saltwater intrusion** results primarily in coastal areas when overdraft lowers the water table. As groundwater removal from these wells continues, saltwater is pulled into the groundwater system making the water no longer usable due to the salt content.

Most groundwater is clean but it can become pollutant or contaminated. Contamination can come from leaking underground storage tanks, leaking landfills, septic systems, accidental spills, and agriculture pesticides, and excess nutrients can also make their way into groundwater systems. The type of contamination can vary from bacteria, metals, hydrocarbons, and nitrates. Excess nitrates in water causes **methemoglobinemia** or **"blue-baby"** syndrome. The consumption of nitrates, especially in children, causes increases in methemoglobin. Methemoglobin blocks the production of hemoglobin and causes oxygen stress in the body. Oxygen stress means the body is not getting enough oxygen resulting in a bluish tint of the skin along with shortness of breath, headaches, and fatigue.

Part C *Depleting Groundwater (Groundwater Mining)*

1. Draw and label the aquifer layers in the groundwater model on the diagram in Figure 14.2. Include the following terms: water table, unconfined aquifer, confined aquifer, vadose zone, and saturated zone.

2. a. What happens if water from the river is overused for agriculture or industry? Write a hypothesis (H_{WM}) that predicts what will happen to the water availability in the wells when water is overused from the River.

    H_{WM}:

3. Observe and mark any change in the water level of the wells, underground tank, pond, and river in the groundwater model as the river is draining. Mark the change on the diagram Figure 14.2 and label this water level as **WL2.**
    a. Does the pond or any of the wells become dry or depleted of water? If so explain why?

    b. Was hypothesis H_{WM} accepted or rejected? Why?

4. What can you do to prevent pollutant entering a well from the surface?

5. What can you do to prevent depletion or overuse of groundwater?

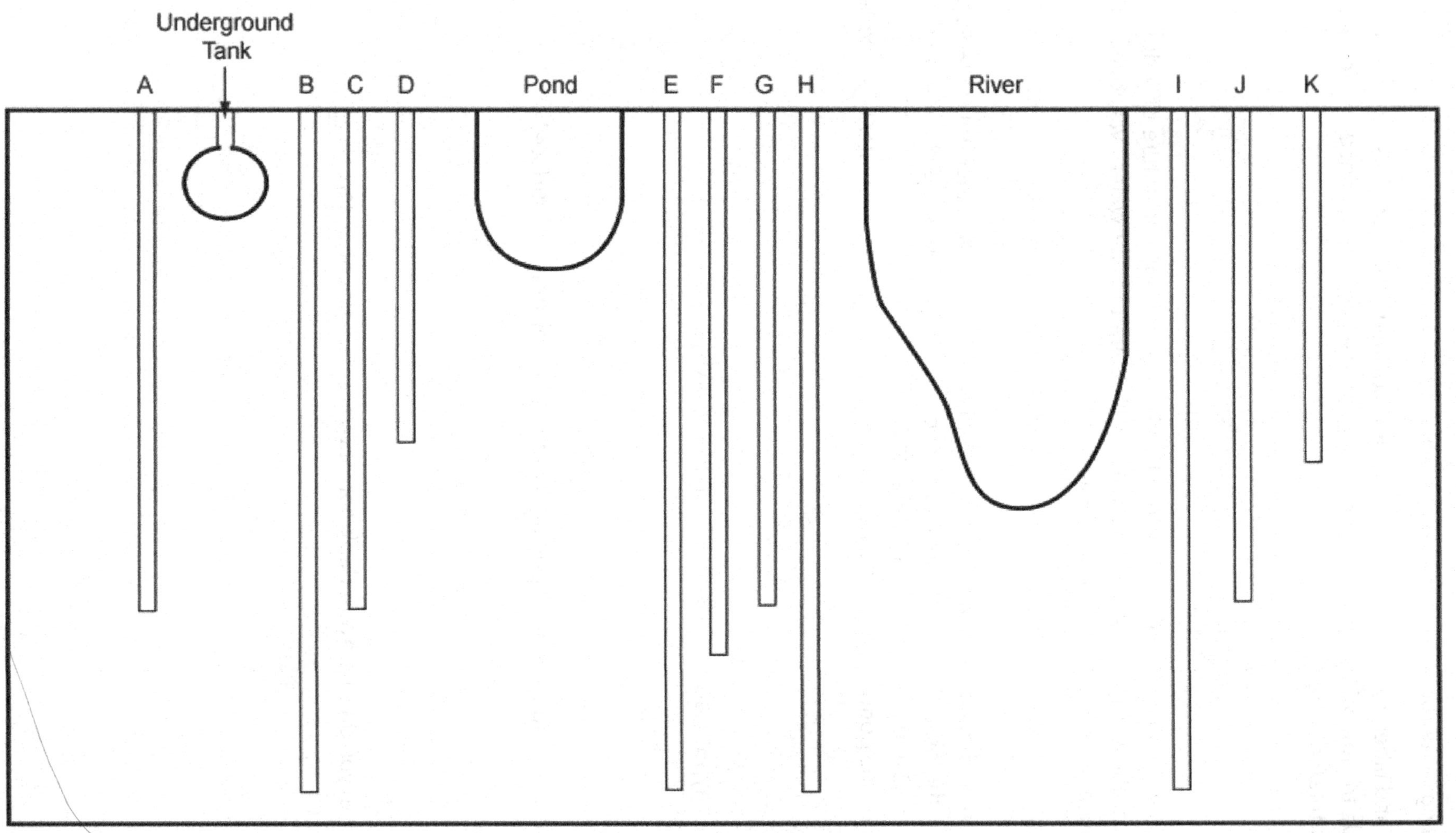

Figure 14.2   Cross section diagram of the groundwater model for Part A. Draw and label aquifer layers.

Wetlands Lab

Objectives

1. Understand the three key criteria for defining wetlands.
2. Understand the function and importance of wetland

Materials Needed

Sponges
Simulation model (rectangle aluminum pans)
Beakers
Sediment
Stir rods
Bottles of runoff water
MicroLab
Vials

Introduction

Historically, wetlands were long considered insect-ridden wastelands. It has been estimated that in the 1600's the United States had 221 million acres of wetlands. As the population in the United State grew, there was expansion into new territory and many wetlands were drained to be used for agriculture. In addition, building of railroads, trails and roads through large wetlands, such as the Ohio Black Swamp, permanently changed the region. By the 1960's there were many financial and institutional incentives to drain or destroy wetlands. The Federal Government encouraged land drainage and wetland destruction through a variety of legislations and policies. Many of these policies included subsidies for public-works projects, agriculture, and flood-control projects built on wetlands. By the 1980's only 103 million acres of wetlands remained in the United States. Although there was an awareness of the importance of wetlands in ecosystem function in the 1970's, it wasn't until 1987 when Federal efforts started the restoration of wetlands. Today wetlands have been recognized as important ecosystems *for flood control, groundwater recharge, water purification, breeding habitat, and habitat for many endangered animals.*

Wetlands are areas where water covers the soil all or part of the year. In addition wetlands contain hydric soil and water tolerant plants (hydrophytes). Wetland types vary greatly because of regional and local difference in soil, topography, climate, hydrology, water chemistry, vegetation and other factors. Two general categories of wetlands are coastal or tidal wetlands and non-tidal wetlands. **Coastal wetlands** include those found along the Atlantic, Pacific, Alaskan and Gulf coast. Many of these are closely linked to where salt and freshwater meet such as in estuaries. These include mud flats, sand flats, mangrove swamps and tidal salt marshes. **Inland or non-tidal wetlands** are common on flood plains, along rivers and streams, along margins of lakes and ponds and in other low-lying areas. These include vernal pools, bogs, potholes, wet meadows and riparian wetlands.

Surface waters can degrade due to inputs of excess nutrients, sediment, metals, hydrocarbons or other pollutants into the water. The origins of these pollutants can be from either point or non-point sources. **Point source pollutants** are a single identifiable source from which pollutants are discharged such as water exiting a waste water treatment facility, factory or paper mill. These types of releases are regulated by federal and state agencies through a National Pollutant Discharge Elimination System (NPDES) permit. **Non-point source** pollution comes from many diffused sources. This source of pollution can results from a wide variety of primarily human activities such as construction, mining, logging, and poor land management. One type of runoff that has received more attention in recent years is stormwater runoff. Stormwater is the water from rain or snow melt that does not soak into the ground but runs off into waterways. This non-point source of water pollution can originate from rooftops, roadways, bare soil or other impervious surfaces. As the water moves over the land or impervious surfaces it accumulates debris, chemicals, and sediment that could adversely affect the water quality if discharged untreated. To handle stormwater, many cities and towns have a **combined sewage system** in which one network of pipes is used for all types of sewage (household, stormwater and industrial). This type of system is less expensive to install, but can be a problem as the population and infrastructure increases. The increase in water and impervious surface areas increases the stormwater runoff and the demand on the city's sewage system. In a combined system, all collected sewage water is passed through a waste water treatment plant. Therefore the system has to be large enough to handle the enlarged inflows from rainstorms to avoid allowing untreated polluted sewage water released into surface waters. It is uneconomical to have these large treatment facilities; so many urban areas use a separate sewer system. With a **separate sewage system**, storm water is collected and release into surface water with little or no treatment. To prevent harmful pollutants from being washed or dumped into the surface water, cities must obtain a NPDES permit and develop a storm water management plan. Federal Environmental Protection Agency as well as the Ohio Environmental Protection agency promotes **best management practice** (BMP) for controlling and treating stormwater runoff to reduce pollutants and to reduce the amount of water discharge into the sewer system. One of these practices is the implementation of **constructed or artificial wetlands**, including rain gardens and retention ponds, for treatment of storm water. Wetlands use a variety of chemical, biological and physical processes to remove the pollutants from the water protecting water quality. Constructed or artificial wetlands mimic the treatment that occurs in natural wetland systems. Artificial wetlands can be used in the treatment of both point sources, such as waste water, and non-point source pollutants, such as in agricultural runoff, to reduce the amount of runoff pollutants that enter a natural water system. These units naturally filter and biodegrade the pollutants from the stormwater runoff water.

Part A. Wetlands Absorbing Runoff

Experiment 1

1. Using a ring stand and clamp, set up the shoe box model so that one end is set to height 2 on the ring stand and the other end is balanced with a 100 mL beaker.
2. Determine the slope of the box (Figure 1.0).
   - To determine the slope measure the vertical distanced traveled by the water (rise), divided by the horizontal distance traveled by the water (run).

treatment systems: tall grass material (felt), wetlands (sponges), short grass material (cloth towel) and mixed vegetation (bumpy liner). The area designated with a black line is the space you have to design a filtration system on the 'simulation model'; your biological treatment system cannot go beyond this black line and has to use two of the four material options. The creek cannot be altered or the border changed; the creek is designated at the one end of the 'simulation model'.

Develop a hypothesis based on your design's ability to reduce water and sediment from storm water runoff.

Hypothesis:

Design Instructions:

1.   As a group, decided which material will be used for your design to reduce storm water runoff; you can only use the material provided.

2.   *Draw* a schematic of your design in Figure 2

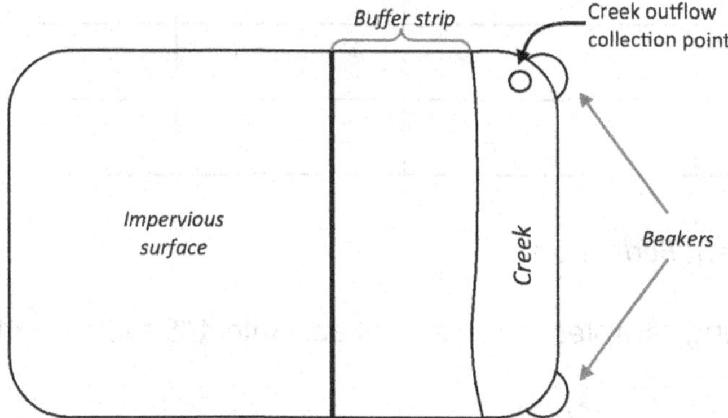

Figure 16.2   Graphic of treatment system for 'simulation model' of storm water runoff.

3.   Obtain the storm water runoff from the lab instructor.

4.   Set the 'simulation model' so that the top of the 'simulation model' is elevated at height 2.

5.   Place a beaker at the creek outflow collection point. Place a second beaker on the opposite corner so that the model will not fall over.

6.   Shake simulated storm water and pour a sample into a spectrometry vial. Cap with an unlabeled cap and set aside for later use.

7.   Measure out 75 mL of storm water runoff.

8.   Pour the water into the small pan with holes on the bottom to simulate rain water.

9.   Allow the water to flow over the impervious surface and collect all the water at the creek outflow point.

10.   Any water that is left in the small pan should be gently poured on to the model.

11. Input the volume of water collected into Table 2.

12. Pour a sample of Simulation Run1 water outflow into a spectrometry vial. Cap the vial using a cap with the label "1". Save for later use.

13. Record observations of the collected runoff water in Table 2.

14. Wring out the design material (sponge, felt, cloth towel) in the sink

15. Repeat the simulated run with new storm water for a total of **3 simulations**. *Be sure to shake the storm water between each use.* At the end of each run, measure the amount of water collected and input results into Table 2. Pour a sample of the outflow from each run into a vial and cap with appropriate labeled cap.

Table 16.2   Results obtains from three simulated precipitation events

Simulation Model	Amt. storm water collected at outflow collection point (mL)	Spectrometry Reading		Description of sediment water
		Relative Conc	Absorp	
Untreated	-------------------	100		
Run 1				
Run 2				
Run 3				

Measuring Relative Sediment Reduction

A. Take vials containing samples to the MicroLab. Select "Spectrometry" and enter a name and hit "OK".

B. There will be a blank vial containing deionized water at each MicroLab station. Put this into the top of the MicroLab and hit "Blank" button on the screen.

C. Invert vial containing untreated storm water 5 times. Set vial into MicroLab and allow it to settle for **1 minute**. While you are waiting, select "Add" insert name to identify the sample (e.g. Untreated or Pretreated) and input a concentration of 100. At 1 minute, hit the "OK" button on the screen.

D. Select the "Absorption" tab from the top of the screen.

E. From the lower set of tabs, select "3 Curve". Select "Linear Fit." This will graph your two points and use this to give relative concentrations for your unknown samples.

F. Select "4 Read" from the lower set of tabs and press "Add". Insert a name (e.g. Run 1) but do not press "OK". Invert sample 5 times and set into MicroLab, **wait 1 minute** then press "OK".

G. Repeat this procedure with Run 2 and 3; input the relative concentration in Table 2.

H. After all readings have been taken, examine the water samples in the vials. In Table 2 describe the amount of heavy sediment found in each vial as well as the clarity of each sample.

Answer the following questions based on your observations.

1. Was your hypothesis accepted or rejected? Why?

2. Did the simulation change between each run? How did the simulation change?

3. What would have happened if the water was not removed from the design material after each run?

4. What natural methods/processes would remove water from a vegetative area (think about the hydrologic cycle)?

5. *Based on your simulations*, what problems could you anticipate having if you were installing this design along a natural creek system?

6. What *other* types of problems (that may not have been tested in this simulation) could interfere with your design or the ability of your design to function properly?

7. What design alterations or recommendations would you make if you were going to run another simulation (time, money, material, etc. are not limited)?

Website Resources:

US EPA Wetland Protection and Restoration: *https://www.epa.gov/wetlands*
Wetlands in National Parks: *https://www.nps.gov/subjects/wetlands/about.htm*
USGS Why are wetlands & aquatic habitats important: *https://water.usgs.gov/edu/qa-around-wetlands.html*